Advanced Structured Materials

Volume 96

Series editors

Common engineering materials reach in many applications their limits and new developments are required to fulfil increasing demands on engineering materials. The performance of materials can be increased by combining different materials to achieve better properties than a single constituent or by shaping the material or constituents in a specific structure. The interaction between material and structure may arise on different length scales, such as micro-, meso- or macroscale, and offers possible applications in quite diverse fields.

This book series addresses the fundamental relationship between materials and their structure on the overall properties (e.g. mechanical, thermal, chemical or magnetic etc.) and applications.

The topics of *Advanced Structured Materials* include but are not limited to

- classical fibre-reinforced composites (e.g. class, carbon or Aramid reinforced plastics)
- metal matrix composites (MMCs)
- micro porous composites
- micro channel materials
- multilayered materials
- cellular materials (e.g. metallic or polymer foams, sponges, hollow sphere structures)
- porous materials
- truss structures
- nanocomposite materials
- biomaterials
- nano porous metals
- concrete
- coated materials
- smart materials

Advanced Structures Material is indexed in Google Scholar and Scopus.

More information about this series at http://www.springer.com/series/8611

Sorin Vlase · Marin Marin
Andreas Öchsner

Eigenvalue and Eigenvector Problems in Applied Mechanics

Springer

Sorin Vlase
Department of Mechanical Engineering
Transilvania University of Braşov
Braşov, Romania

Marin Marin
Department of Mathematics
and Computer Science
Transilvania University of Braşov
Braşov, Romania

Andreas Öchsner
Fakultät Maschinenbau
Esslingen University of Applied Sciences
Esslingen, Germany

ISSN 1869-8433 ISSN 1869-8441 (electronic)
Advanced Structured Materials
ISBN 978-3-030-00990-8 ISBN 978-3-030-00991-5 (eBook)
https://doi.org/10.1007/978-3-030-00991-5

Library of Congress Control Number: 2018957055

This Springer imprint is published by the registered company Springer Nature Switzerland AG
The registered company address is: Gewerbestrasse 11, 6330 Cham, Switzerland

Preface

This volume presents, in a unitary way, some problems of applied mechanics, analyzed using the matrix theory and the properties of eigenvalues and eigenvectors. Problems and situations of different nature are studied. Different problems and studies in mechanical engineering lead to patterns that are treated in a similar way. The same mathematical apparatus allows the study of mathematical structures such as the quadratic forms but also mechanical problems such as multibody rigid mechanics, continuum mechanics, vibrations, elastic and dynamic stability or dynamic systems. A substantial number of engineering applications illustrate this volume.

Braşov, Romania — Sorin Vlase
Braşov, Romania — Marin Marin
Esslingen, Germany — Andreas Öchsner

Contents

Chapter 1
Vectors

1.1 Vectors. Fundamental Notions

Mechanics uses physical quantities that cannot be characterized only by their magnitude (how it happens with a scalar) but also need other attributes to define them, namely the direction, or point of application (fixed or mobile). They represent a multitude of quantities, frequently used such as forces acting on a material point or solid, speeds, accelerations, moment of force etc. These quantities are called *vectors*.

A vector contains the measure of the magnitude or length (arithmetic element) to which the direction (geometric element) is added. The direction is defined with one word as the vector orientation.

It follows that the vector is geometrically represented by an oriented segment. The direction on which the vector will act is given by the line segment, with a definite direction (it is a binary dimension geometrically indicated by an arrow at the end of the segment) and the magnitude (positive numerical value) is given by the length of the segment, which is represented by a certain convenient scale. In Fig. 1.1 a vector as an oriented segment is represented. The point A is called the initial point or origin, base, tail and the point B is the terminal (final) point, head, endpoint, or tip (see Liesen and Mehrmann 2015; Springer 2013; Simionescu 1982).

The vector notations are different, depending on the geometric or algebraic approaches as well as the authors. Several notations have come to terms and are presented below. Because the vector is geometrically represented by a directed line segment, it has an origin called, for example, A and one end called, for example, B. In this case, the vector is marked with an arrow. The first letter indicates the origin and the other the end. If the extremities are not called, vectors can be denoted by an arrow letter $\left(\vec{a}, \vec{b}, \vec{u}, \vec{r}\right)$ or by a bar letter on the top $\left(\bar{a}, \bar{b}, \bar{u}, \bar{r}\right)$. Sometimes the bar is dropped and the letter is written in lowercase boldface as: $(\mathbf{a}, \mathbf{b}, \mathbf{u}, \mathbf{r})$ or lowercase italic boldface as: $(\boldsymbol{a}, \boldsymbol{b}, \boldsymbol{u}, \boldsymbol{r})$, especially in the Anglo-Saxon literature, in more mathematical works. If algebraic representations are used, the vectors are also written

S. Vlase et al., *Eigenvalue and Eigenvector Problems in Applied Mechanics*, Advanced Structured Materials 96, https://doi.org/10.1007/978-3-030-00991-5_1

Fig. 1.1 A vector $\overline{AB}$

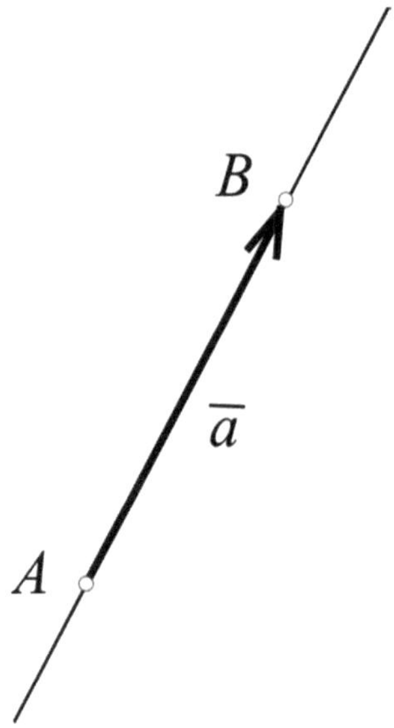

as column matrix: $\{a\}$, $\{b\}$, $\{u\}$, $\{r\}$. In contrast to vectors, matrices are denoted by uppercase letters.

The vector magnitude (length or intensity) is denoted as in the case of algebra. Thus, the vector magnitude of $\overline{AB}$ is $\left|\overrightarrow{AB}\right|$, for the vectors $\vec{a}, \vec{b}, \vec{u}, \vec{r}$, the modules will be $|\vec{a}|$, $\left|\vec{b}\right|$, $|\vec{u}|$, $|\vec{r}|$ or simpler as a, b, u, r. Sometimes for **a**, **b**, **u**, **r**, the notation $\|\mathbf{a}\|$, $\|\mathbf{b}\|$, $\|\mathbf{u}\|$, $\|\mathbf{r}\|$, is used.

Classification of vectors: in mechanics, to characterize a vector used in a certain type of problem, additional data is required. Thus, the vectors were classified into the following three classes:

- a vector with fixed origin and head is called a bound vector. A fixed vector is the moment of a force in a point.
- a vector is called a free vector when only the direction and magnitude matter and the origin is of no importance; so it can be considered any point of space.

For example velocity or acceleration.

- sliding vectors are vectors whose point of application can be found anywhere on the support line of the vector without the mechanical effect on the body changing. Sliding vectors are forces that act on a rigid or angular velocity vector.

Equality of vectors:

Definition Two vectors are equal if they have the same magnitude and direction, they can be placed on the same or parallel lines.

The equality is expressed by the algebraic sign "=". For example: $\vec{a} = \vec{b}$ or $\bar{a} = \bar{b}$ or $\{a\} = \{b\}$.

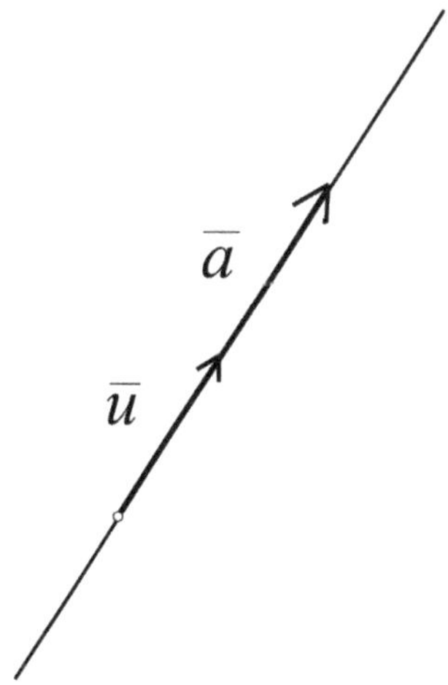

Fig. 1.2 Representation of the unit vector

Equal vectors located on parallel lines are called equipollent vectors and the operation by which a vector is moved by a translation on a line parallel to the vector support is called an equipollent operation.

If two vectors have the same magnitude but opposite directions, they are called opposite vectors and this is expressed by: $\bar{a} = -\bar{b}$.

Two vectors that do not have the same magnitude but have the same support line direction are called collinear vectors, regardless of the direction.

Unit vector: If we give a vector $\bar{a}$, we construct another vector $\bar{u}$ with the same direction as $\bar{a}$, whose magnitude is equal to unit size: $u=1$ (see Fig. 1.2). The vector $\bar{u}$ is called the unit vector. If this choice is made, any colinear vector with $\bar{a}$ can be expressed with $\bar{b} = \lambda\bar{u}$.

Axis: If on a straight line we chose a positive sense, an origin and a unit of length, then we have defined an axis. The vector of the axis is the unit vector located on it, the sense of which coincides with the positive direction of the axis.

1.2 Vector Operations

1.2.1 Addition of Vectors

1.2.1.1 Sum of Two Vectors

Two vectors $\bar{a}$ and $\bar{b}$ are considered, see Fig. 1.3. We move both vectors through an equipollency in a point O of the space. It obtains a parallelogram formed with these two vectors. The diagonal of this parallelogram is called the sum of the vectors $\bar{a}$ and

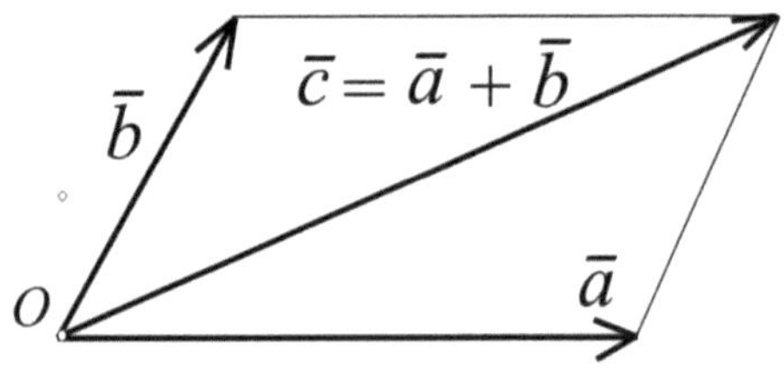

Fig. 1.3 Addition of two vectors

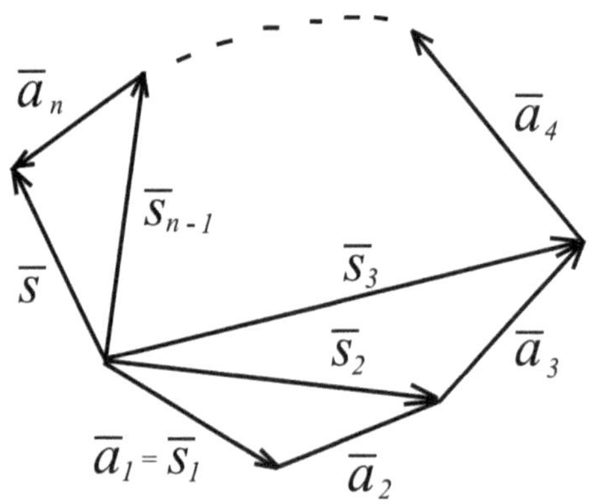

Fig. 1.4 Sum of several vectors

$\bar{b}$. The vectors $\bar{a}$ and $\bar{b}$ are called components of the vector $\bar{c}$ and the vector $\bar{c} = \bar{a} + \bar{b}$ is called the resultant vector of the given vectors.

The geometric operation by which the resultant vector is constructed is named the parallelogram rule. The rule of addition has an experimental basis and is considered an axiom if the vectors are forces. From a geometric point of view, the vector $\bar{c}$ can be obtained by placing at the head of vector $\bar{a}$ the origin point of vector $\bar{b}$. Then, vector $\bar{c}$ resulting as the third side of the triangle so formed. If the rule of addition is considered, then it is immediately seen that it can be written: $\bar{a} + \bar{b} = \bar{b} + \bar{a}$ that is, the addition of two vectors is a commutative operation.

1.2.1.2 Sum of Several Vectors

The sum can be naturally generalized if we deal with multiple vectors. Consider the vectors $\bar{a}_1, \bar{a}_2, \ldots, \bar{a}_n$. With vectors $\bar{a}_1$ and $\bar{a}_2$ we construct the sum vector $\bar{s}_2 = \bar{a}_1 + \bar{a}_2$, transporting $\bar{a}_2$ to the head of vector $\bar{a}_1$.

Then we construct the sum vector $\bar{s}_3 = \bar{s}_2 + \bar{a}_1 + \bar{a}_2 + \bar{a}_3$, transporting $\bar{a}_3$ to the endpoint of $\bar{s}_2$ (see Fig. 1.4).

The vector $\bar{s}_n$ can be obtained by mathematical induction as the sum:

$$\bar{s}_n = \bar{a}_1 + \bar{a}_2 + \cdots + \bar{a}_n.$$

Consequently, it results that the sum vector is obtained by constructing the polygon formed by the vectors $\bar{a}_1, \bar{a}_2, \ldots, \bar{a}_n$.

The resultant vector $\bar{s} = \bar{s}_n = \bar{a}_1, \bar{a}_2 + \cdots + \bar{a}_n$, is obtained as an oriented segment with the origin point as the origin point of $\bar{a}_1$ and the end point as the end point of $\bar{a}_n$. This process is called the polygonal contour rule.

For n vectors to have a null sum, the polygon built with them must be closed. In particular, in the case of three vectors, the condition that they have a null sum is that, by placing each vector with the origin at the end of the previous one, a triangle is formed. Because the triangle is a plane figure, the three vectors must be coplanar.

In the case of the sum of two vectors $\bar{c} = \bar{a} + \bar{b}$, since the three form a triangle, one can write the inequality of the triangle in its known form:

$$c \leq a + b. \tag{1.1}$$

In the case of a polygonal contour generated by n vectors, which are placed one at the extremity of the other, this inequality can be generalized in the following form:

$$s = |\bar{a}_1 + \bar{a}_2 + \cdots + \bar{a}_n| \leq a_1 + a_2 + \cdots + a_n. \tag{1.2}$$

We have this equality only if all the vectors are collinear and have the same physical meaning.

1.2.1.3 Properties of Vectors Addition

The sum of the vectors has the following properties:

- The vector addition is commutative. This was shown when the sum was defined. It follows that the order in which the summation of more vectors is ordered is indifferent, that is, we have (Fig. 1.5):

$$\bar{a} + \bar{b} + \bar{c} = \bar{b} + \bar{a} + \bar{c} = \bar{b} + \bar{c} + \bar{a} = \bar{c} + \bar{b} + \bar{a} = \bar{c} + \bar{b} + \bar{a}. \tag{1.3}$$

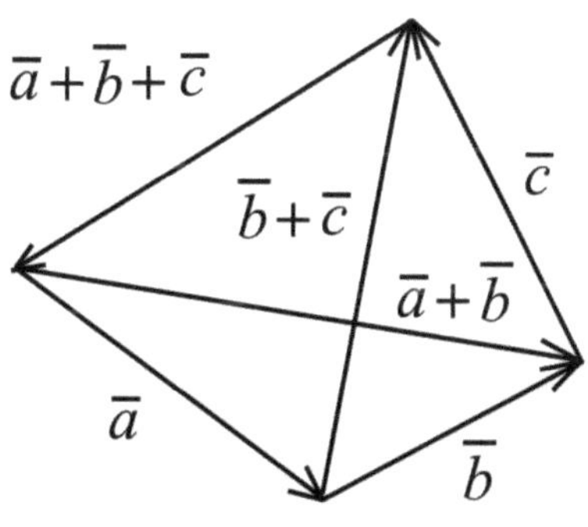

Fig. 1.5 Associativity of the vector addition

- The vector addition is associative. If this is represented graphically, we immediately find that we have:

$$\left(\bar{a}+\bar{b}\right)+\bar{c}=\bar{a}+\left(\bar{b}+\bar{c}\right). \tag{1.4}$$

- The vector addition is distributive over the multiplication with a scalar.

We will consider the vectors $\bar{a}$, $\bar{b}$, $\bar{c}$, $\bar{d}$, their sum is $\bar{s}=\bar{a}+\bar{b}+\bar{c}+\bar{d}$ (see Fig. 1.6) and the sum of the vectors $m\bar{a}$, $m\bar{b}$, $m\bar{c}$ and $m\bar{d}$ is $\bar{s}'=m\bar{a}+m\bar{b}+m\bar{c}+m\bar{d}$. The sum $\bar{s}$ is obtained by constructing the polygonal line *OABCD* with the given vectors, the sum being $\overline{OD}$.

We build the vectors:

$$\overline{OA'}=m\overline{OA}=m\bar{a},\ \overline{OB'}=m\overline{OB}=m\bar{b},$$
$$\overline{OC'}=m\overline{OC}=m\bar{c},\ \overline{OD'}=m\overline{OD}=m\bar{d}.$$

The polygons *OABCD* and *O′A′B′C′D* are similar, having the sides parallel and the same proportion of the sides, so

$$\overline{OD'}=m\overline{OD}=m(\bar{a}+\bar{b}+\bar{c}+\bar{d}).$$

However, we have:

$$\overline{OD'}=\overline{OA'}+\overline{A'B'}+\overline{B'C'}+\overline{C'D'}$$
$$=m\bar{a}+m\bar{b}+m\vec{c}+m\bar{d}$$

By comparing the two expressions for $\overrightarrow{OD'}$ it results:

$$m(\bar{a}+\bar{b}+\bar{c}+\bar{d})=m\bar{a}+m\overleftarrow{b}+m\bar{c}+m\bar{d} \tag{1.5}$$

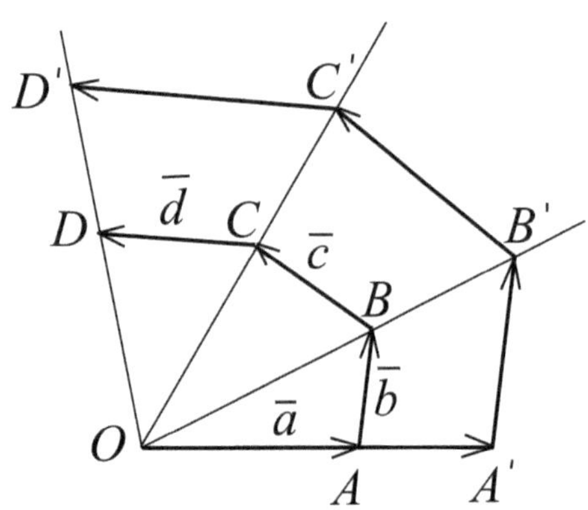

Fig. 1.6 Addition of three vectors

1.2.1.4 Decomposing a Vector

(a) *Two-directional decomposition into two directions that determine a plane parallel to the vector*

Let us consider two straight lines, i.e. Δ_1 and Δ_2. Let O be the origin of the vector $\vec{a}$ and A its head. We move the two straight lines in the origin O using a parallel translation. In this case, based on the mentioned hypotheses, the two lines and the given vector will be in the same plane (see Fig. 1.7).

It follows that the problem will be reduced in this case to finding two vectors with the origin in O, $\vec{a}_1$ along the straight line and $\vec{a}_2$ along the straight line (Δ_2) so that their sum is the given vector $\vec{a}$.

Taking into account that the rule of addition of two vectors is given by the parallelogram rule, we will have to construct a parallelogram, with the sides on the lines (Δ_1) and (Δ_2) which will have as diagonal the vector $\vec{a}$. For this through the point A, representing the head of the vector $\vec{a}$, one needs to construct two straight lines parallel to (Δ_1) and (Δ_2). One obtains the parallelogram OA_1AA_2, whose OA_1 and OA_2 sides represent exactly the vectors we are looking for, i.e. $\vec{a}_1$ and $\vec{a}_2$.

If the two straight lines (Δ_1) and (Δ_2) are perpendicular then $\vec{a}_1$ and $\vec{a}_2$ represent the orthogonal projections of vector $\vec{a}$ on the two straight lines in both directions.

The decomposition of a vector in more than two directions, located in the same plan, is undetermined (that is, there is an infinite number of decompositions of a vector in three directions in the plane). This can be simply demonstrated if one arbitrary component is taken in one direction and the other component is decomposed in two directions.

(b) *Decomposing a vector in three non-parallel directions in space*

We note with Ox, Oy and Oz the three concurrent directions in which we want to decompose the vector $\bar{a}$ with the origin in O. If the three directions are somewhere in the space, we can move the vectors through O, using a parallel translation.

To perform the decomposition we proceed as follows: the lines OA and OA_3 determine a plane whose intersection with the Oxy plane is the straight line Ot. Now the vector $\vec{a}$ can be decomposed along the directions Oz and Ot directions so that: $\bar{a} = \overline{OA_3} + \overline{OB} = \bar{a}_3 + \overline{OB}$. Furthermore, the $\overline{OB}$ vector found in the Oxy plane can

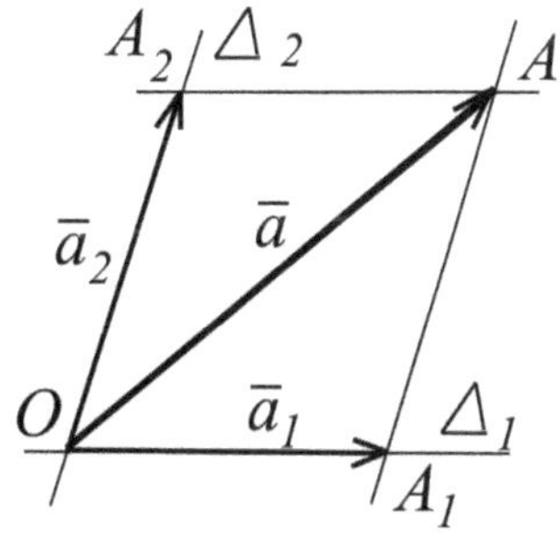

Fig. 1.7 Decomposition of a vector

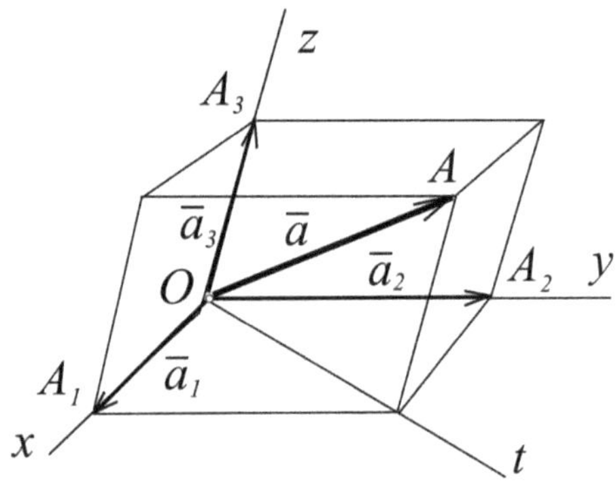

Fig. 1.8 Decomposition of a vector in three directions

be decomposed in the directions Ox and Oy so that: $\overline{OB} = \overline{OA_1} + \overline{OA_2} = \bar{a}_1 + \bar{a}_2$. It follows that the vector $\bar{a}$ can be uniquely decomposed in the three directions given so that (Fig. 1.8):

$$\bar{a} = \overline{OA_3} + \overline{OA_1} + \overline{OA_2} = \bar{a}_3 + \bar{a}_1 + \bar{a}_2.$$

If the three directions are the coordinate axes of an orthogonal system, then the three projections are called the orthogonal projections of the vector on the three axes.

If the components $\bar{a}_1, \bar{a}_2, \bar{a}_3$ are considered the sides of a parallelepiped then the given vector is the diagonal to this parallelepiped.

The decomposition of a vector in more than three directions not founded in the same plane is indetermined.

(c) *Algebraic representation of vectors*

In the following, the concepts of vectorial calculus will be presented in both geometric and algebraic representation, since applications can be approached sometimes more easily in the geometric representation and sometimes in the algebraic representation.

Thus, if three unit vectors $\bar{i}, \bar{j}, \bar{k}$ of the same origin are considered, perpendicular two by two, which indicate the directions of the axes Ox, Oy, Oz, they will form an orthogonal coordinate system (see Fig. 1.9).

If a certain vector $\bar{a}$, is considered this vector can be decomposed in the three axes and it can be written as:

$$\bar{a} = a_x\bar{i} + a_y\bar{j} + a_z\bar{k}. \tag{1.6}$$

The magnitudes a_x, a_y, a_z are named by the components of vector $\bar{a}$ in the orthogonal coordinate system $Oxyz$. We mention here that it is possible to use a not-orthogonal system, but in the following, because the orthogonal system is used almost exclusively in common applications, the reference will be made to it, unless otherwise specified.

Fig. 1.9 Components of a vector in an orthogonal coordinate system

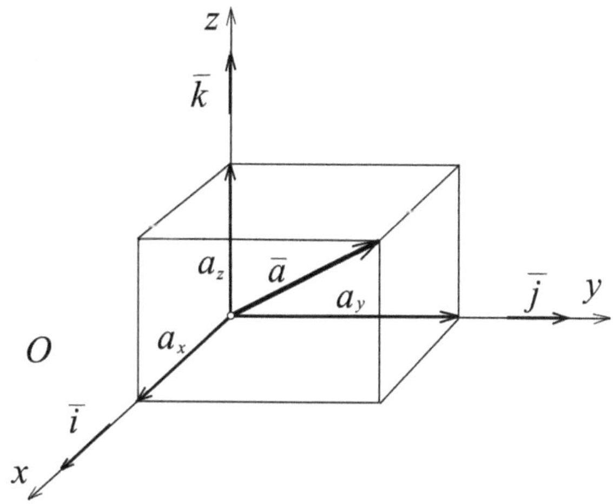

In this case the given vector can also be written in the form $\bar{a}(a_x, a_y, a_z)$ or

$$\{a\} = \begin{Bmatrix} a_x \\ a_y \\ a_z \end{Bmatrix}, \tag{1.7}$$

in order to indicate its components. In this case the sum of two vectors can be defined as follows: if $\bar{c} = \bar{a} + \bar{b}$ where $\bar{a}(a_x, a_y, a_z)$ and $\bar{b}(b_x, b_y, b_z)$ then the vector $\bar{c}$ will be the vector having the components:

$$c_x = a_x + b_x; \quad c_y = a_y + b_y; \quad c_z = a_z + b_z \quad .$$

The same can be written in matrix form as:

$$\{c\} = \begin{Bmatrix} c_x \\ c_y \\ c_z \end{Bmatrix} = \{a\} + \{b\} = \begin{Bmatrix} a_x \\ a_y \\ a_z \end{Bmatrix} + \begin{Bmatrix} b_x \\ b_y \\ b_z \end{Bmatrix} = \begin{Bmatrix} a_x + b_x \\ a_y + b_y \\ a_z + b_z \end{Bmatrix}. \tag{1.8}$$

With this definition of the sum of two vectors, which will be reduced to a sum of real numbers, all the properties of the sum shown above are easily demonstrated.

1.2.1.5 Subtraction of Two Vectors

Definition. To subtract the vector $\bar{b}$ from the vector $\bar{a}$, the task is to find the vector $\bar{c}$ that added to $\bar{b}$ gives $\bar{a} : \bar{b} + \bar{c} = \bar{a}$. Subtracting of vectors is denoted by a minus sign "−":

$$\bar{c} = \bar{a} - \bar{b}. \tag{1.9}$$

Geometrically, the subtraction can be made in two ways:

- for the parallelogram with a diagonal a and a side b at which the angle between a and b is known, we will have to construct the side c of the parallelogram, which is joining the extremities of the vectors b and a and then leading through the head of a parallel to b and leading through the origin parallel to c. The direction of c is obtained from the condition that b summed with c must give a. It can be written as follows: $\bar{c} = \overline{BA} = \overline{OA} - \overline{OB}$, (the relation of Chasles), i.e. the difference vector has as origin the extremity of the first vector and as end point the end point of the second vector.
- it can be written $\bar{c} = \bar{a} - \bar{b} = \bar{a} + (-\bar{b})$, that is, it adds $\bar{a}$ to the vector $(-\bar{b})$ (see Fig. 1.10).

Algebraically, the subtraction can be written:

$$\{c\} = \begin{Bmatrix} c_x \\ c_y \\ c_z \end{Bmatrix} = \{a\} - \{b\} = \begin{Bmatrix} a_x \\ a_y \\ a_z \end{Bmatrix} - \begin{Bmatrix} b_x \\ b_y \\ b_z \end{Bmatrix} = \begin{Bmatrix} a_x - b_x \\ a_y - b_y \\ a_z - b_z \end{Bmatrix}, \tag{1.10}$$

i.e. the components of the difference are equal to the difference of the components of the two vectors.

1.2.1.6 Some Properties of Addition and Subtraction

- If the equality $\bar{a} = \bar{b}$ is given, then a vector $\bar{c}$ can be added to both sides and the equality remains.
- Given the equality $\bar{a} = \bar{b}$, the same vector $\bar{c}$ can be subtracted from both sides and the equality remains. From the two properties it results that in a vectorial equality

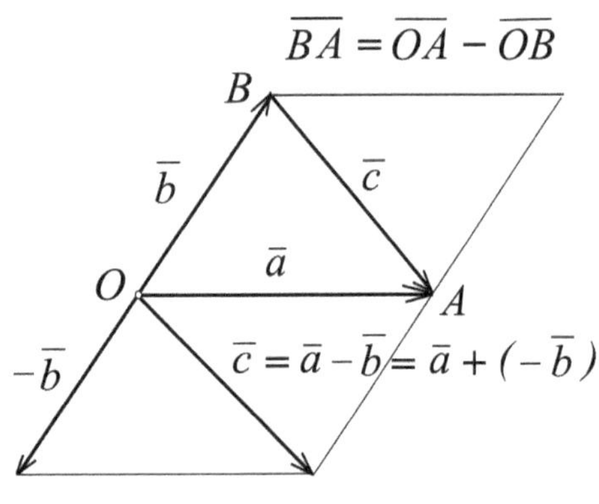

Fig. 1.10 Subtraction of two vectors

some terms can be passed from one part of the equality to the other, with a changed sign, as in any algebraic equality.

- If we have the vector equalities $\bar{a}_1 = \bar{b}_1$ and $\bar{a}_2 = \bar{b}_2$, then they can be added or subtracted, term by term, and the equality remains.

1.2.1.7 Decomposing a Vector in Two Vectors

To decompose a known vector $\bar{a}$ after two vectors $\bar{b}$ and $\bar{c}$ returns to decompose the vector $\bar{a}$ in the directions of $\bar{b}$ and $\bar{c}$; it can be written:

$$\overline{OA} = \overline{OA_1} + \overline{OA_2}. \tag{1.11}$$

These components can be expressed according to the vectors $\bar{b}$ and $\bar{c}$ (see Fig. 1.11) after the relations: $\overline{OA_1} = m\bar{b}$ and $\overline{OA_2} = n\bar{c}$ since the components are vectors collinear with vectors $\bar{b}$ and $\bar{c}$. It results:

$$\bar{a} = m\bar{b} + n\bar{c}. \tag{1.11$'$}$$

It follows that any vector can be decomposed in two coplanar vectors with it. If the two vectors are not collinear, decomposition is unique. If the vectors are collinear decomposition is indefinite.

The written relationship can be brought in a homogeneous form if we take:

$$m = -\frac{\beta}{\alpha}; \quad n = -\frac{\gamma}{\alpha}.$$

In this case the relation (1.11) can be written as:

$$\alpha\bar{a} + \beta\bar{b} + \gamma\bar{c} = 0, \tag{1.12}$$

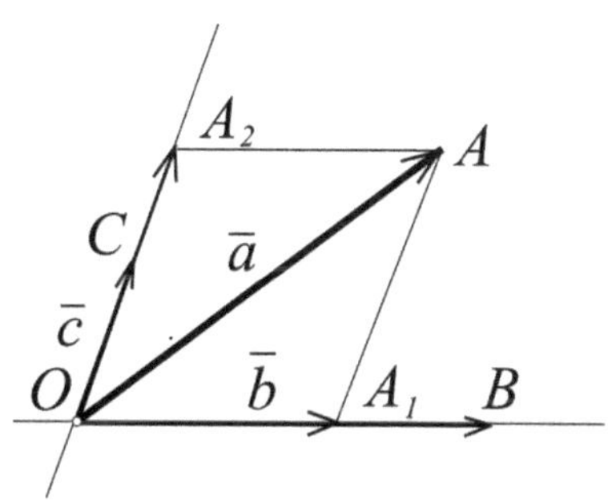

Fig. 1.11 Decomposition of a vector in two vectors

which is the condition that three vectors lie in the same plane. The relationships shows that if three coplanar vectors are given, any one of them can be expressed as a linear combination of the other two.

Algebraically, the relation (1.12) can be written as:

$$\begin{aligned} \alpha a_x + \beta b_x + \gamma c_x &= 0, \\ \alpha a_y + \beta b_y + \gamma c_y &= 0, \\ \alpha a_y + \beta b_y + \gamma c_y &= 0. \end{aligned} \tag{1.13}$$

According to the theory of homogeneous linear systems, there are non-zero α, β, γ that satisfy the system if and only if the system determinant is zero:

$$\begin{vmatrix} a_x & b_x & c_x \\ a_y & b_y & c_y \\ a_z & b_z & c_z \end{vmatrix} = 0, \tag{1.14}$$

which is the condition that three vectors are in the same plane (one column is a linear combination of the other two).

Theorem *The decomposition of a given vector in two other vectors, coplanar with it, is unique.*

If it is assumed that there would be two decompositions we would have: $\bar{a} = m\bar{b} + n\bar{c}$ *and* $\bar{a} = m'\bar{b} + n'\bar{c}$ *with* $m \neq m'$ *and* $n \neq n'$. *If we subtract these two relationships, we get* $(m - m')\bar{b} + (n - n')\bar{c} = 0$ *or* $\bar{b} = -\frac{(n-n')}{(m-m')}\bar{c} = \lambda\bar{c}$ *with* $\lambda = -\frac{(n-n')}{(m-m')}$, *that is,* $\bar{b}$ *and* $\bar{c}$ *are collinear, which contradicts the hypothesis. It is very easy to show that the cases* $m = m'; n \neq n'$ *and* $m \neq m'; n = n'$ *also contradict the hypothesis, and it only remains:* $m = m'; n = n'$ *that is, the decomposition is unique.*

Theorem *The necessary and sufficient condition for three vectors to be coplanar is to have the relationship (1.11) or (1.12) between them.*

If the coplanar vectors $\bar{a}$, $\bar{b}$ *and* $\bar{c}$ *are considered then* $\bar{a}$ *can be decomposed in the other two vectors and it can be written:* $\bar{a} = m\bar{b} + n\bar{c}$ *(the condition is necessary).*

Reciprocally, the previously written relationship shows that $\bar{a}$ *is the sum of two vectors collinear with* $\bar{b}$, *respectively* $\bar{c}$, *that is a coplanar vector with them (the condition is sufficient).*

1.2.1.8 Scalar Multiplication

Let us consider a vector $\bar{a}$ and the scalar λ. By definition, the product of a vector with a scalar λ is a vector $\bar{b}$, colinear with $\bar{a}$, has the same direction with $\bar{b}$ if λ is positive and opposite direction if λ is negative and has the magnitude equal to $|\lambda|\bar{a}$. It is written as:

$$\vec{b} = \lambda\vec{a}. \tag{1.15}$$

1.2.1.9 Linearly Independent and Linearly Dependent Coplanar Vectors

Since between two non-collinear vectors $\bar{a}$ and $\bar{b}$ there cannot be a linear relationship (if $\bar{a} = \lambda\bar{b}$ or $\alpha\bar{a} + \beta\bar{b} = 0$ the vectors would be collinear, i.e. a contradiction to the hypothesis), it is said that two non-collinear vectors are linearly independent.

Let us now consider three vectors that are in the same plane. Since relations (1.11) or (1.12) hold between them, there is a linear dependence between them. For this reason, they are called linearly dependent.

If two non-collinear vectors $\bar{b}$ and $\bar{c}$ are considered, any coplanar vector with them can be expressed as (1.11) or (1.12). So using these two vectors can express any other coplanar vector with them. These vectors are called fundamental vectors.

1.2.1.10 Decomposing a Vector in Three Non-coplanar Vectors

We consider a vector $\bar{a}$ and we propose to write it as the sum of three collinear vectors respectively with $\bar{b}$, $\bar{c}$, $\bar{d}$ (they are not coplanar). For this, the three vectors go through an equipollency into a common O-point of the space, where are the tail of the vector $\bar{a}$. It decomposes $\bar{a}$ along the $\bar{b}$, $\bar{c}$, $\bar{d}$ directions and it contains the components as (Fig. 1.12):

$$\bar{a} = \overline{OA_1} + \overline{OA_2} + \overline{OA_3} \tag{1.16}$$

Since $\overline{OA_1}, \overline{OA_2}, \overline{OA_3}$ are collinear respectively with $\bar{b}$, $\bar{c}$, $\bar{d}$, it can be written: $\overline{OA_1} = m\bar{b}$, $\overline{OA_1} = n\bar{c}$, $\overline{OA_1} = p\bar{d}$

Then it follows:

$$\bar{a} = m\bar{b} + n\bar{c} + p\bar{d}. \tag{1.16'}$$

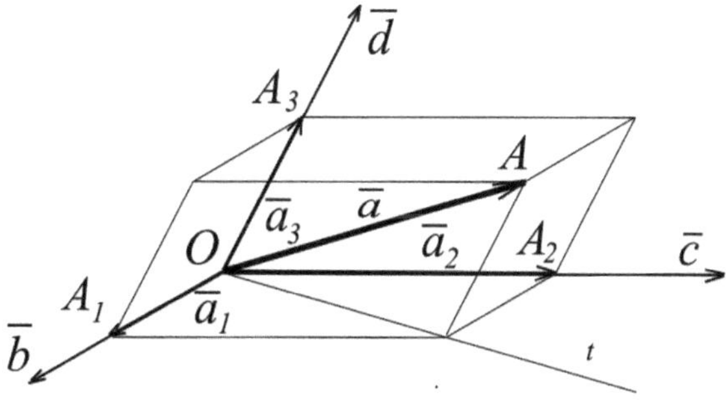

Fig. 1.12 Decomposition of a vector after three vectors

If we introduce the following notations:

$$m - \beta/\alpha, n = -\gamma/\alpha, p = -\delta/\alpha,$$

we obtain the following expression:

$$\alpha\bar{a} + \beta\bar{b} + \gamma\bar{c} + \delta\bar{d} = 0. \tag{1.17}$$

Theorem *The decomposition of a vector in three non-coplanar vectors is unique.*

Proof It is assumed that there are two distinct decompositions:

$$\bar{a} = m\bar{b} + n\bar{c} + p\bar{d} \quad \text{and} \quad \bar{a} = m'\bar{b} + n'\bar{c} + p'\bar{d}.$$

By subtraction it follows:

$$0 = (m - m')\bar{b} + (n - n')\bar{c} + (p - p')\bar{d}. \tag{1.18}$$

If $m \neq m', n \neq n', p \neq p'$ it follows, according to the previous relationship, that the three vectors $\bar{b}, \bar{c}, \bar{d}$ are coplanar, in contradiction with the hypothesis. If $m = m', n \neq n', p \neq p'$, according to Eq. (1.17), it will result:

$$0 = (n - n')\bar{c} + (p - p')\bar{d}, \tag{1.18$'$}$$

that means that $\bar{c}$ and $\bar{d}$ are collinear, which is also contrary to the hypothesis. Cases in which only $n = n'$ or only $p = p'$ are treated analogously and lead to a violation of the hypotheses. It follows that only the case $m = m', n = n', p = p'$ and the decomposition is unique.

1.2.1.11 Linearly Independent and Linearly Dependent Non-coplanar Vectors

If we have three non-coplanar vectors $\bar{b}, \bar{c}, \bar{d}$, then there cannot be a linear relation of the form: $\beta\bar{b} + \gamma\bar{c} + \delta\bar{d} = 0$ because this relationship characterizes three coplanar vectors. It is said that the three non-coplanar vectors are linearly independent. If we have a set of four vectors, there will be the relationship (1.15) or (1.16) between them which shows that one of them is a linear combination of the other three. For this reason, they are called linearly dependent.

If a group of three non-coplanar (linearly independent) vectors are chosen, it is possible to express any vector from the three-dimensional space with them. That is why these vectors are called the fundamental vectors.

1.2.1.12 The Conditions in Which Three Vectors Have Heads on a Line or Four Vectors Which Have the Heads in the Same Plane

We give, without demonstration, two theorems that define the conditions in which three vectors which have the heads (terminal points) on a straight line or four vectors have the heads in the same plane.

Theorem I *The necessary and sufficient condition for the terminal point of the vector* $\overline{OC} = \bar{c}$ *to be on the straight line* AB *is that, in the relationship* $\bar{d} = m\bar{a} + n\bar{b} + p\bar{c}$*, we have* $m + n + p = 1$.

Theorem II *The necessary and sufficient condition for the vector's* $\overline{OD} = \bar{d}$ *head to be found in the plane* ABC *is that, in the relationship* $\bar{c} = m\bar{a} + n\bar{b}$*, we have* $m + n = 1$.

1.2.2 *Dot Product*

1.2.2.1 Definition and Properties

Consider two vectors $\bar{a}$ and $\bar{b}$ and we define an operation between these two vectors resulting in the space of the real numbers, as follows:

Definition The number resulting from the product of the magnitude of the two vectors multiplied by the cosine of the angle between them is called the dot product of the two vectors $\bar{a}$ and $\bar{b}$.

The dot product is marked with a point: $c = \bar{a} \cdot \bar{b} = ab\cos\left(\bar{a}, \bar{b}\right)$. The scalar product has the following properties:

- The dot product is null if: (a) one of the vectors is null or (b) if the two vectors are perpendicular. Demonstration of the first property is immediate. For the second demonstration, considering the formula of the cosine of the angle between the two vectors, since the vectors are perpendicular, this cosine is zero and therefore the result is zero.
- The square of a vector relative to the dot product operation is equal to the square of magnitude. If the definition formula is applied and the angle between the two factors of the product is zero, then the result is immediately obtained:

$$\bar{a}^2 = \bar{a} \cdot \bar{a} = a \cdot a\cos 0 = a^2 \quad \left(= |a|^2\right). \tag{1.19}$$

- The dot product is commutative. The demonstration follows immediately from the scalar product definition formula.
- The dot product is distributive over the scalar multiplication. The demonstration is simple if it is taken into account that $\bar{a}$ and $\lambda\bar{a}$ are collinear and $\bar{b}$ together $\lambda\bar{b}$ are also collinear. Thus,

$$\cos(\bar{a}, b) = \cos(\lambda\bar{a}, \bar{b}) = \cos(a, \lambda\bar{b}).$$

Then, from the definition formulas, it simply follows:

$$\lambda(\bar{a} \cdot \bar{b}) = (\lambda\bar{a} \cdot \bar{b}) = (a \cdot \lambda\bar{b}). \tag{1.20}$$

- The dot product of two vectors is equal with the magnitude of one, multiplied by the projection of the other in the direction of the first vector. The property results as outlined in Fig. 1.13. We have:

$$\bar{a}\bar{b} = ab\cos(\bar{a}, \bar{b}) = a\lfloor b\cos(\bar{a}, \bar{b})\rfloor = a \; pr_{\bar{a}}\bar{b}, \tag{1.21}$$

because: $pr_{\bar{a}}\bar{b} = b\cos(\bar{a}, \bar{b})$. Similarly:

$$\bar{a}\bar{b} = b \; pr_{\bar{b}}\bar{a}. \tag{1.22}$$

- We give, without demonstration, the following result: the scalar product is distributive over the vector addition. We have:

$$\bar{a}(\bar{b} + \bar{c}) = \bar{a}\bar{b} + \bar{a}\bar{c}. \tag{1.23}$$

- If the previous property is taken into account, it can be written immediately:

$$(\bar{a} + \bar{b}) \cdot (\bar{c} + \bar{d}) = \bar{a}\bar{c} + \bar{b}\bar{c} + \bar{a}\bar{d} + \bar{b}\bar{d},$$

  i.e. the rule of multiplication of two polynomials is preserved.
- The dot product of two vectors does not change if to one of the vectors is added a vector perpendicular to the other. Let us consider two vectors $\bar{a}$ and $\bar{b}$ and a third vector $\bar{u}$ perpendicular to $\bar{a}$. So we have: $\bar{a}\bar{u} = 0$. Then: $\bar{a}(b + \bar{u}) = \bar{a}\bar{b} + \bar{a}\bar{u} = \bar{a}\bar{b}$ what demonstrates the property.
- Next, consider an orthogonal coordinate system $Oxyz$ that has the unit vectors $\bar{i}, \bar{j}, \bar{k}$. There are the following relationships, particularly important in the algebraic representation of vectors:

$$\bar{i} \cdot \bar{j} = \bar{j} \cdot \bar{i} = \cos\frac{\pi}{2} = 0, \bar{j} \cdot \bar{k} = \bar{k} \cdot \bar{j} = 0, \bar{k} \cdot \bar{i} = \bar{i} \cdot \bar{k} = 0$$

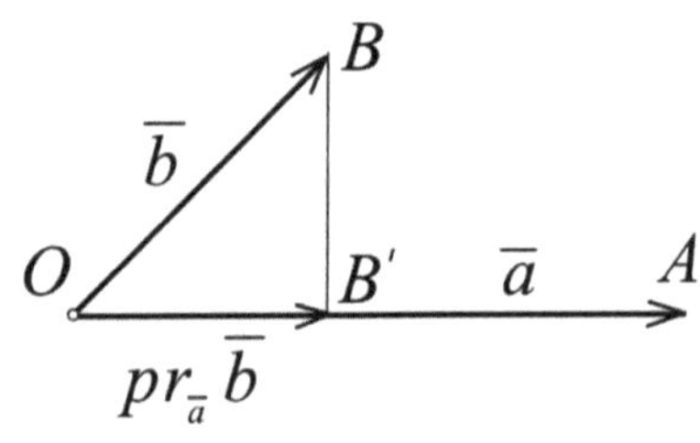

Fig. 1.13 Projection of a vector on another vector

$$\left(\bar{i}\right)^2 = \left(\bar{j}\right)^2 = \left(\bar{k}\right)^2 = \cos 0 = 1. \tag{1.24}$$

1.2.2.2 Algebraic Representations

If the representation (1.6) of vectors $\bar{a}$ and $\bar{b}$ is considered, then it can be written:

$$\begin{aligned}
\bar{a}\cdot\vec{b} &= \left(a_x\bar{i}+a_y\bar{j}+a_z\bar{k}\right)\left(b_x\bar{i}+b_y\bar{j}+b_z\bar{k}\right)\\
&= a_xb_x\left(\bar{i}\right)^2 + a_xb_y\left(\bar{i}\cdot\bar{j}\right) + a_xb_z\left(\bar{i}\cdot\bar{k}\right) + a_yb_x\left(\bar{j}\cdot\bar{i}\right) + a_yb_y\left(\bar{j}\right)^2\\
&\quad + a_yb_z\left(\bar{j}\cdot\bar{k}\right) + a_zb_x\left(\bar{k}\cdot\bar{i}\right) + a_zb_y\left(\bar{k}\cdot\bar{j}\right) + a_zb_z\left(\bar{k}\right)^2\\
&= a_xb_x + a_yb_y + a_zb_z,
\end{aligned} \tag{1.25}$$

i.e. the dot product is given by the sum of the product of the corresponding components of the two vectors. In matrix notation we have:

$$\bar{a}\cdot\bar{b} = \{a\}^T\{b\} = \begin{bmatrix} a_x & a_y & a_z \end{bmatrix}\begin{Bmatrix} b_x \\ b_y \\ b_z \end{Bmatrix} = a_xb_x + a_yb_y + a_zb_z.$$

The condition that two vectors are perpendicular will be written, if this representation is adopted, as:

$$a_xb_x + a_yb_y + a_zb_z = 0. \tag{1.26}$$

The magnitude of a vector is obtained if it is considered the dot product of a vector with itself, from the relation:

$$\bar{a}\cdot\bar{a} = (\bar{a})^2 = a^2 = a_x^2 + a_y^2 + a_z^2, \tag{1.27}$$

from where:

$$a = \sqrt{a_x^2 + a_y^2 + a_z^2}. \tag{1.28}$$

In matrix form, we have:

$$a^2 = \{a\}^T\{a\} = a_x^2 + a_y^2 + a_z^2.$$

The angle between two vectors is obtained from the relationship:

$$\cos\left(\bar{a},\bar{b}\right) = \frac{\bar{a}\cdot\bar{b}}{ab} = \frac{a_xb_x + a_yb_y + a_zb_z}{\sqrt{a_x^2 + a_y^2 + a_z^2}\sqrt{b_x^2 + b_y^2 + b_z^2}}. \tag{1.29}$$

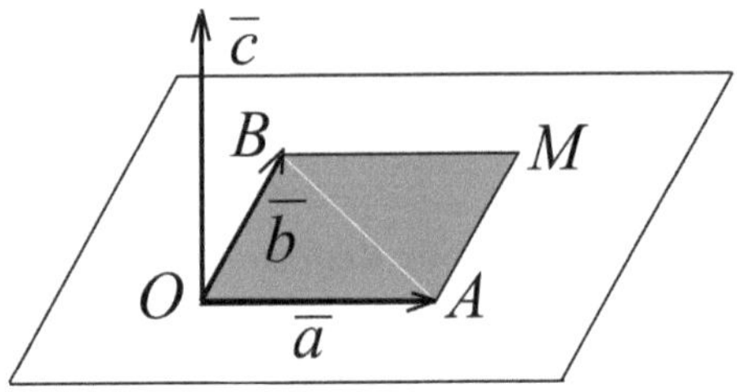

Fig. 1.14 Cross product definition

The unit vector associated with a direction. It is considered a straight line (Δ) that makes the angles α, β and γ with the three axes of an orthogonal coordinate system.

In addition, consider a vector oriented along this line. Its three components will be equal to his projections on the three axes, namely:

$$u_x = u\cos\alpha, \quad u_y = u\cos\beta, \quad u_z = u\cos\gamma.$$

Let us now consider the unit vector of the straight line (Δ), denoted by $\bar{u}_o$. This unit vector is obtained by dividing the vector by its length:

$$\bar{u}_o = \frac{\bar{u}}{u} = \bar{i}\cos\alpha + \bar{j}\cos\beta + \bar{k}\cos\gamma. \tag{1.30}$$

In matrix form, we can write:

$$\{u_o\} = \begin{Bmatrix} \cos\alpha \\ \cos\beta \\ \cos\gamma \end{Bmatrix}. \tag{1.31}$$

Because $\bar{u}_o$ is an unit vector, it results immediately:

$$\cos^2\alpha + \cos^2\beta + \cos^2\gamma = 1. \tag{1.32}$$

1.2.3 Cross Product

1.2.3.1 Definition and Properties

Consider two vectors and define an operation between the two vectors having as result a vector, called the cross product (or vector product or outer product or directed area product) (Fig. 1.14).

Definition Given the vectors $\bar{a} = \overrightarrow{OA}$, $\bar{b} = \overrightarrow{OB}$ we define an operation, called the cross product of the two vectors, which associates a vector $\bar{c} = \overrightarrow{OC}$ determined in the following way:

(a) it is perpendicular to the plane determined by the given vectors (this determines its direction);
(b) the direction of the vector is given by the right-hand rule;
(c) the magnitude of the vector $\bar{c} = \overrightarrow{OC}$ is equal to the area of the parallelogram that vectors $\overrightarrow{OA}$ and $\overrightarrow{OB}$ span.

So it follows $c = ab\sin\left(\bar{a}, \bar{b}\right) = 2\ area(OAB)$. The vector product is marked with the "×" sign: $\bar{c} = \bar{a} \times \bar{b}$.

Here are some of the properties of the vector product:

- The cross product is null if one of the two vectors is null or if the two vectors have the same direction. Indeed, if in the expression of the cross product magnitude it is considered that the angle between the two vectors is zero, then the sine of this vector is zero, so the magnitude of the vector is zero. In particular, the cross product of a vector with itself is zero.
  Parallelism of two vectors. Two vectors will be parallel if their cross product is zero. This is obvious if it is taken into account that the angle between the two parallel vectors is zero, so the sine of this angle, which appears in the formula of the magnitude of the cross product, is zero. This property allows the testing of situations where two vectors are parallel.
- The cross product is anticommutative. Let us consider the products $\bar{a} \times \bar{b}$ and $\bar{b} \times \bar{a}$. The resulting vector in the two cases has the same direction, being oriented by a straight line perpendicular to the two vectors, but the orientation will be different. If we take $\bar{a}$ over $\bar{b}$ over the shortest way, we get an orientation, and if we take $\bar{b}$ over $\bar{a}$ over the shortest way we get an opposite orientation. Since the magnitudes of the two vectors are equal, it can be written that:

$$\bar{a} \times \bar{b} = -\bar{b} \times \bar{a} \quad \text{or} \quad \bar{a} \times \bar{b} + \bar{b} \times \bar{a} = 0. \tag{1.33}$$

- The cross product is distributive in regards to a multiplication with a scalar. The demonstration is simple if it is taken into account that $\bar{a}$ and $\lambda\bar{a}$ are collinear and, in the same time, $\bar{b}$ and $\lambda\bar{b}$ are collinear. In this case, the angle between $\bar{a}$ and $\bar{b}$ as the angle between $\lambda\ \bar{a}$ and $\bar{b}$ or $\bar{a}$ and $\lambda\ \bar{b}$ is the same, so the sine of this angle will be the same in all three cases. It results:

$$\lambda(\bar{a} \times \bar{b}) = (\lambda\bar{a}) \times \bar{b} = \bar{a} \times (\lambda\bar{b}). \tag{1.34}$$

- The cross product remains unchanged when the head point of a vector moves on a parallel line to the other vector (see Fig. 1.15). Let us consider the parallelogram *OABM* generated by the vectors $\bar{a}$ and $\bar{b}$ and the parallelogram *OApBpMp* generated by the vectors $\bar{a}$ and $\bar{b}_p$. The vector $\bar{b}_p$ was obtained by moving its end point along a line parallel to $\bar{a}$. The two parallegrams have the same area because they have the same base and the same height. It follows that the cross product magnitude in the two cases is the same. Since the direction and sense does not change by moving

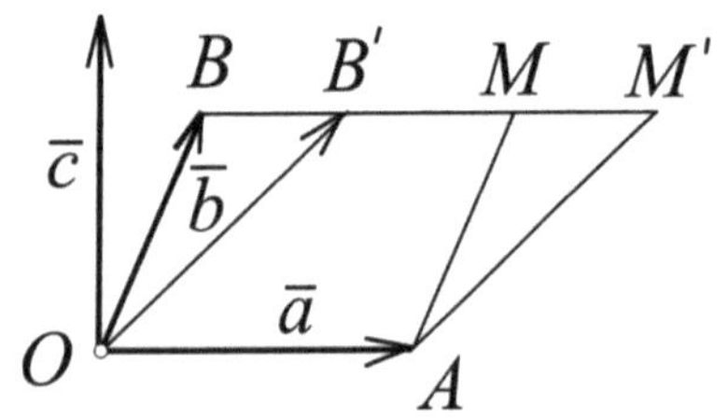

Fig. 1.15 Cross product is invariant if B moves along a line parallel to $\bar{a}$

the head point of $\bar{b}$ along a parallel line with $\bar{a}$ it results that the cross product vector is the same in the two cases.

- The cross product is distributive in regards to vector addition. Because the demonstration, without being difficult, is longer, we do not give it. Thus, it can be written:

$$\bar{a} \times \left(\bar{b} + \bar{c}\right) = \bar{a} \times \bar{b} + \bar{a} \times \bar{c}. \tag{1.35}$$

- Using the previous property, we can write the cross product of two sums of vectors as:

$$\left(\bar{a} + \bar{b}\right) \times \left(\bar{c} + \bar{d}\right) = \bar{a} \times \bar{c} + \bar{a} \times \bar{d} + \bar{b} \times \bar{c} + \bar{b} \times \bar{d}. \tag{1.35'}$$

The rule of multiplication remains the same as for polynomials, with the difference that the order of the factors in the product is not arbitrary, but in each term of development the first factor must belong to the first brackets and the second factor must belong to the second brackets. It can be shown as:

$$\bar{a} \times \bar{b} + \bar{c} \times \bar{d} = (\bar{a} - \bar{c}) \times \left(\bar{b} - \bar{d}\right) + \bar{a} \times \bar{d} + \bar{c} \times \bar{b}. \tag{1.35''}$$

- The cross product does not change if we add to one of the vectors another vector parallel to the second vector of the product. The proof is trivial.
- If an orthogonal trihedral with the unit vectors $\bar{i}$, $\bar{j}$, $\bar{k}$ is considered, taking into account the angles between the coordinate axes and the direction of the resultant vectors, it can be written:

$$\bar{i} \times \bar{i} = \bar{j} \times \bar{j} = \bar{k} \times \bar{k} = 0,$$

and:

$$\bar{i} \times \bar{j} = \bar{k}, \quad \bar{j} \times \bar{k} = \bar{i}, \quad \bar{k} \times \bar{i} = \bar{j}. \tag{1.36}$$

A coordinate system for which the above relations are valid is called the right orthogonal coordinate system. If on the third axis the direction is defined by the relationship $\bar{i} \times \bar{j} = -\bar{k}$ we say that we have a left orthogonal coordinate system.

Generally, the right orthogonal coordinate system will still be used unless otherwise specified.

1.2.3.2 Algebraic Representations

Let us now consider two vectors $\bar{a}$ and $\bar{b}$ defined by their components:

$$\bar{a} = a_x\bar{i} + a_y\bar{j} + a_z\bar{k}, \quad \bar{b} = b_x\bar{i} + b_y\bar{j} + b_z\bar{k}.$$

Taking into account the rules of multiplication of the unit vectors, it can be written:

$$\begin{aligned}
\bar{c} = \bar{a} \times \bar{b} &= \left(a_x\bar{i} + a_y\bar{j} + a_z\bar{k}\right) \times \left(b_x\bar{i} + b_y\bar{j} + b_z\bar{k}\right)\\
&= a_xb_x\left(\bar{i} \times \bar{i}\right) + a_xb_y\left(\bar{i} \times \bar{j}\right) + a_xb_z\left(\bar{i} \times \bar{k}\right) + a_yb_x\left(\bar{j} \times \bar{i}\right) + a_yb_y\left(\bar{j} \times \bar{j}\right)\\
&\quad + a_yb_z\left(\bar{j} \times \bar{k}\right) + a_zb_x\left(\bar{k} \times \bar{i}\right) + a_zb_y\left(\bar{k} \times \bar{j}\right) + a_zb_z\left(\bar{k} \times \bar{k}\right)\\
&= \left(a_yb_z - a_zb_y\right)\bar{i} + \left(a_zb_x - a_xb_z\right)\bar{j} + \left(a_xb_y - a_yb_x\right)\bar{k}.
\end{aligned}$$

Thus, the components of the vector will be given by the expressions:

$$c_x = a_yb_z - a_zb_y; \quad c_y = a_zb_x - a_xb_z; \quad c_z = a_xb_y - a_yb_x. \tag{1.37}$$

Symbolically, the cross product can be represented by the determinant:

$$\bar{c} = \bar{a} \times \bar{b} = \begin{vmatrix} \bar{i} & \bar{j} & \bar{k} \\ a_x & a_y & a_z \\ b_x & b_y & b_z \end{vmatrix}.$$

In matrix form, for the representation of the cross product, the skew symmetric 3×3 matrix, associated with the vector $\bar{a}$ is:

$$[a] = \begin{bmatrix} 0 & -a_z & a_y \\ a_z & 0 & -a_x \\ -a_y & a_x & 0 \end{bmatrix},$$

and then the cross product is represented by the matrix product:

$$[a]\{b\} = \begin{bmatrix} 0 & -a_z & a_y \\ a_z & 0 & -a_x \\ -a_y & a_x & 0 \end{bmatrix} \begin{Bmatrix} b_x \\ b_y \\ b_z \end{Bmatrix} = \begin{Bmatrix} a_yb_z - a_zb_y \\ a_zb_x - a_xb_z \\ a_xb_y - a_yb_x \end{Bmatrix}. \tag{1.38}$$

1.2.4 Scalar Triple Product

1.2.4.1 Definition and Properties

Three vectors are considered, i.e. $\bar{a}$, $\bar{b}$ and $\bar{c}$. The scalar triple product of these three vectors is a scalar, defined by the combination:

$$d = \bar{a} \cdot \left(\bar{b} \times \bar{c}\right). \tag{1.39}$$

The triple scalar product has an interesting geometric significance. Consider the three vectors, $\bar{a}, \bar{b}$ and $\bar{c}$. They will form the parallelepiped $OBDCA'B'D'C'$, see Fig. 1.16.

The magnitude of the cross product $\bar{b} \times \bar{c}$, as shown above, represents the area of the parallelogram $OBDC$, and the cross product is the vector OM, perpendicular to the plane determined by the vectors $\bar{b}$ and $\bar{c}$. Let us now take the perpendicular from A' to OM. $A'A''$ is perpendicular to OM and will result that OA'' is the height of the parallelepiped. Then the dot product between $\overrightarrow{OM}$ and $\bar{a}$ will have a magnitude equal to the product of OM and its projection of $\bar{a}$ onto OM, i.e., OA'' (a value equal to the product between the parallelogram base area and the height of the parallelepiped), so it will be equal to the parallelepiped volume. The scalar triple product may be positive or negative depending on the orientation of the trihedral formed with the three given vectors. If the orientation of this trihedral is positive, the scalar triple product is positive and if the orientation is negative, the scalar triple product is negative.

As a result, the scalar triple product of three vectors represents, in absolute value, the volume of the parallelepiped built on these vectors considered as edges.

Scalar triple product properties

- The scalar triple product of three vectors remains unchanged if its factors are circularly permutated, i.e. we have:

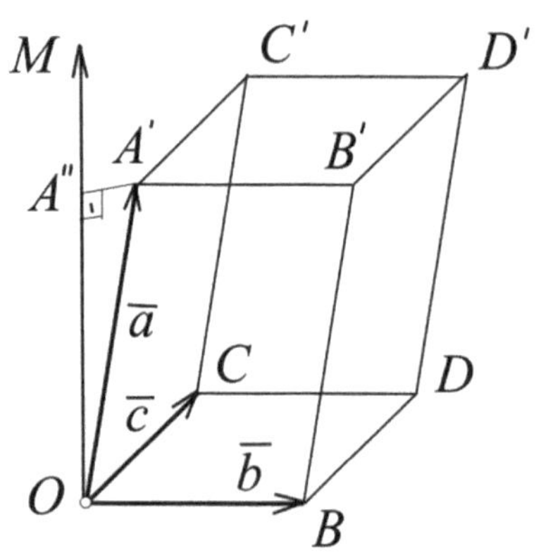

Fig. 1.16 Graphical representation of the scalar triple product

$$\bar{a}(\bar{b} \times \bar{c}) = \bar{b}(\bar{c} \times \bar{a}) = \bar{c}(\bar{a} \times \bar{b}). \tag{1.40}$$

The result can be easily demonstrated if it is taken into account that the three products represent the volume of the parallelepiped built with the three vectors as edges, and the sign is the same in the three cases as the orientation of the trihedral is maintained.

- The scalar triple product does not change if the signs and x are allowed between them, i.e.

$$\bar{a}(\bar{b} \times \bar{c}) = (\bar{a} \times \bar{b})c. \tag{1.41}$$

The result is simply demonstrated by taking into account that the scalar product is commutative and whether the above-mentioned property (which says that the mixed product does not change if the circular permutations are made), is used. Since the value of the scalar triple product depends only on the three vectors and the orientation of the trihedral defined by them, regardless of the sign that is placed between them or x, it is agreed to note the mixed product as:

$$\bar{a}(\bar{b} \times \bar{c}) = (\bar{a} \times \bar{b})c = (\bar{a}\bar{b}\bar{c}) = [\bar{a}\bar{b}\bar{c}]. \tag{1.42}$$

- If the two factors of the triple scalar product change between them, it changes the sign of the result:

$$(\bar{a}\bar{b}\bar{c}) = -(\bar{b}\bar{a}\bar{c}) = -(\bar{a}\bar{c}\bar{b}) = -(\bar{c}\bar{b}\bar{a}). \tag{1.43}$$

- The scalar triple product is null if one of the factors is null or if the three vectors are coplanar. The first case is obvious. For the second case it will be noticed that if the three vectors are coplanar, the parallelepiped built with them has a zero volume, so the mixed product is null. A particular case, commonly found in practice, is when two vectors are collinear, so the three vectors are coplanar. So, if any two vectors of the mixed product are collinear, the mixed product is null.

1.2.4.2 Algebraic Representations

If the vectors $\bar{a} = a_x\bar{i} + a_y\bar{j} + a_z\bar{k}$, $\bar{b} = b_x\bar{i} + b_y\bar{j} + b_z\bar{k}$ and $\bar{c} = c_x\bar{i} + c_y\bar{j} + c_z\bar{k}$ are considered, and the calculus are made, for the expression of the mixed product, one obtains:

$$(\bar{a}\bar{b}\bar{c}) = a_x b_y c_z + a_y b_z c_x + a_z b_x c_y - a_z b_y c_x - a_y b_x c_z - a_x b_z c_y. \tag{1.44}$$

This expression is precisely the determinant of the matrix having the vector components $\bar{a}$, $\bar{b}$ and $\bar{c}$ as lines, and so it can be written:

$$(\bar{a}\bar{b}\bar{c}) = \begin{vmatrix} a_x & a_y & a_z \\ b_x & b_y & b_x \\ c_x & c_y & c_z \end{vmatrix}. \tag{1.45}$$

If we write the scalar triple product in this form, all of the above-mentioned properties are immediately demonstrated, if the determinant properties are taken into account.

1.2.5 Vector Triple Product

If the vectors $\bar{a}$, $\bar{b}$ and $\bar{c}$ are considered the vector triple product is the vector $\bar{d}$:

$$\bar{d} = \bar{a} \times (\bar{b} \times \bar{c}). \tag{1.46}$$

In the following we give, without demonstration (which does not pose problems but is more laborious), the main result of the vector triple product. Thus, the vector triple product $\bar{d}$ can be expanded as:

$$\bar{d} = \bar{a} \times (\bar{b} \times \bar{c}) = (\bar{a}\bar{c})\bar{b} - (\bar{a}\bar{b})\bar{c}. \tag{1.47}$$

The vector triple product is null if one of the factors is null, if the vectors $\bar{b}$ and $\bar{c}$ are collinear or if the vector $\bar{a}$ is perpendicular to the plane determined by the vectors $\bar{b}$ and $\bar{c}$. The demonstration is immediate.

1.2.6 Applications of Vector Calculus

1.2.6.1 Position Vector of a Point

Let us consider a point M in space and choose a fixed point O as the origin of the space. Let us build the vector $\vec{r} = \overrightarrow{OM}$. The point of M is completely determined by the vector and thus constructed. The vector is called the position point of the vector M. In applications, different coordinate systems (cartesian, cylindrical, polar, spherical, natural etc.) can be used to define using different scalar components. These components unambiguously define the position vector, thus the position of the point M.

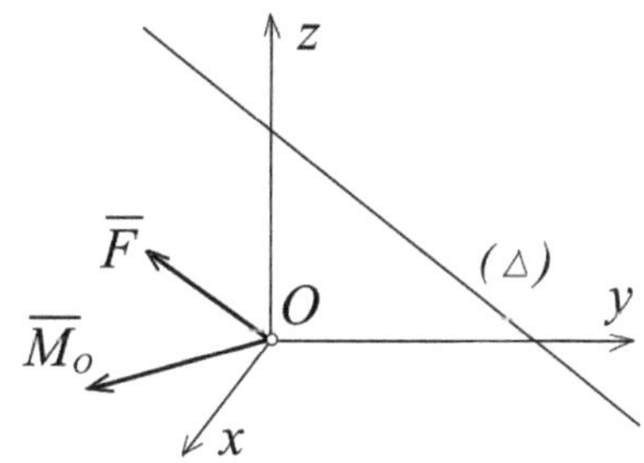

Fig. 1.17 Support line of a force

1.2.6.2 Support Line of a Force

Let us consider the force $\vec{F}$ and its moment $\vec{M}_O$ in the point O (see Fig. 1.17), (we have $\vec{F}\vec{M}_O = 0$). We propose to determine the support line of the force. For this, we need to determine the solution $\vec{r}$ from the vector equation $\vec{r} \times \vec{F} = \vec{M}_O$. Writing n component, the relationship will be:

$$\begin{aligned} yZ - zY &= M_{Ox}, \\ zX - xZ &= M_{Oy}, \\ xY - yX &= M_{Oz}, \end{aligned}$$

or

$$\begin{bmatrix} 0 & Z & -Y \\ -X & 0 & X \\ Y & -Z & 0 \end{bmatrix} \begin{Bmatrix} x \\ y \\ z \end{Bmatrix} = \begin{Bmatrix} M_{Ox} \\ M_{Oy} \\ M_{Oz} \end{Bmatrix}.$$

It is easy to see that the system determinant is zero:

$$\begin{vmatrix} 0 & Z & -Y \\ -X & 0 & X \\ Y & -Z & 0 \end{vmatrix} = 0. \tag{1.48}$$

Thus, the system does not have a single solution. If we consider the first two equations as the main equations, it is obtained that the characteristic determinant is equal to zero:

$$\Delta_c = \begin{vmatrix} 0 & Z & M_{Ox} \\ -X & 0 & M_{Oy} \\ Y & -Z & M_{Oz} \end{vmatrix} = Z\left(XM_{Ox} + YM_{Oy} + ZM_{Oz}\right) = Z\left(\vec{R}\vec{M}_O\right) = 0,$$

by virtue of the fact that moment and force are two perpendicular vectors. It follows that we have an undetermined system of two equations with three unknowns. Geometrically seen, the two equations are planes whose intersection will give us a line, which is the support line of the force.

Another method of determining the force support line is a vector method. Thus, if we premultiply the vector equation:

$$\vec{r} \times \vec{F} = \vec{M}_O, \tag{1.49}$$

to left, with $\vec{F}$, in a cross product, one obtains:

$$\vec{F} \times \left(\vec{r} \times \vec{F}\right) = \vec{F} \times \vec{M}_O,$$

and expanding the vector triple product it results:

$$F^2\vec{r} - \left(\vec{F}\vec{r}\right)\vec{F} = \vec{F} \times \vec{M}_O,$$

and:

$$\vec{r} = \frac{\vec{F} \times \vec{M}_O}{\vec{F}^2} + \frac{\left(\vec{F}\vec{r}\right)}{F^2}\vec{F}. \tag{1.50}$$

Since on the basis of previous considerations we have seen that not all components of the vector are independent, one can choose as a parameter the expression:

$$\lambda = \frac{\left(\vec{F}\vec{r}\right)}{F^2},$$

and we obtain:

$$\vec{r} = \frac{\vec{F} \times \vec{M}_O}{F^2} + \lambda\vec{F}. \tag{1.51}$$

The equation obtained is a line, which has the direction of the force and passes through the point defined by the vector:

$$\vec{d} = \frac{\vec{F} \times \vec{M}_O}{F^2}. \tag{1.52}$$

The vector $\vec{d}$ represents the distance from the origin to the line (it is perpendicular to the force—it comes from a cross product—so too on the support line and in addition when $\lambda = 0$ it will result that its end belongs to the line). So the force support line has the equation:

$$\vec{r} = \vec{d} + \lambda \vec{F}, \tag{1.53}$$

or, in components:

$$\begin{aligned} x &= d_x + \lambda X, \\ y &= d_y + \lambda Y, \\ z &= d_z + \lambda Z. \end{aligned} \tag{1.54}$$

By removing the parameter λ, the line can also be expressed as:

$$\frac{x - d_x}{X} = \frac{y - d_y}{Y} = \frac{z - d_z}{Z}. \tag{1.55}$$

1.3 Applications

1.3.1. If the vectors $\bar{a}$, $\bar{b}$ and $\bar{c}$ are given, demonstrate that the following relationship exists:

$$\bar{a} \times \left(\bar{b} \times \bar{c}\right) + \bar{b} \times (\bar{c} \times \bar{a}) + \bar{c} \times \left(\bar{a} \times \bar{b}\right) = 0.$$

1.3.2. If we have four vectors $\bar{a}$, $\bar{b}$, $\bar{c}$ and $\bar{d}$ show that we have the following relationship:

$$\left(\bar{a} \times \bar{b}\right) \cdot (\bar{c} \times \bar{d}) = \begin{vmatrix} \bar{a} \cdot \bar{c} & \bar{a} \cdot \bar{d} \\ \bar{b} \cdot \bar{c} & \bar{b} \cdot \bar{d} \end{vmatrix}.$$

1.3.3. If we have the vectors $\bar{a}$, $\bar{b}$, $\bar{c}$ and $\bar{d}$ show that we have the following relationship:

$$\left(\bar{a} \times \bar{b}\right) \times \left(\bar{c} \times \bar{d}\right) = \left(\bar{a}\bar{b}\bar{d}\right)\bar{c} - \left(\bar{a}\bar{b}\bar{c}\right)\bar{d} = \left(\bar{c}\bar{d}\bar{a}\right)\bar{b} - \left(\bar{c}\bar{d}\bar{b}\right)\bar{a}.$$

1.3.4. Either vectors $\bar{a}$, $\bar{b}$ and $\bar{c}$ are non-coplanar. Let it show that if:

$$\left(\bar{a} \times \bar{b}\right) \cdot \left(\bar{c} \times \bar{d}\right) = \left(\bar{a} \cdot \bar{b}\right) \cdot \left(\bar{b} \cdot \bar{c}\right),$$

then the vectors $\bar{a}$ and $\bar{c}$ are perpendicular.

1.3.5. Show that:

$$\left(\bar{a} \times \bar{b}\right) \times \left(\bar{b} \times \bar{c}\right) = \left(\bar{a}\bar{b}\bar{c}\right)\bar{b}.$$

1.3.6. Show that:

$$\left[(\bar{a}\times\bar{b})(\bar{b}\times\bar{c})(\bar{c}\times\bar{a})\right]=(\bar{a}\bar{b}\bar{c})^2.$$

1.3.7. Show that we have the expansion:

$$(\bar{a}\times\bar{b})\times\left[(\bar{b}\times\bar{c})\times(\bar{c}\times\bar{a})\right]=(\bar{a}\bar{b}\bar{c})\left[(\bar{c}\bar{a})\bar{b}-(\bar{c}\bar{b})\bar{a}\right].$$

1.3.8. If $a_i, b_i, c_i (i = 1, 2, 3)$ are real numbers, show that the vectors:

$$\begin{aligned}\bar{u}_1&=(\bar{b}_1-\bar{c}_1)\bar{i}+(c_1-a_1)\bar{j}+(a_1-b_1)\bar{k}\\\bar{u}_2&=(\bar{b}_2-\bar{c}_2)\bar{i}+(c_2-a_2)\bar{j}+(a_2-b_2)\bar{k}\\\bar{u}_3&=(\bar{b}_3-\bar{c}_3)\bar{i}+(c_3-a_3)\bar{j}+(a_3-b_3)\bar{k},\end{aligned}$$

are coplanar.
Note: If the mixed product of the three vectors is written as a determinant, it is found to be zero, so the vectors are coplanar.

1.3.9. If the non-coplanar vectors $\bar{a}, \bar{b}$ and $\bar{c}$ are given, calculate the scalar triple product of the vectors: $(\bar{a}+\bar{b}), (\bar{b}+\bar{c}), (\bar{c}+\bar{a})$ and interpret geometrically the result.

$$Solution: P=2(\bar{a}\bar{b}\bar{c})$$

1.3.10. If the non-coplanar vectors $\bar{a}, \bar{b}$ and $\bar{c}$ are given, calculate the scalar triple product of the vectors: $(\bar{a}-\bar{b}), (\bar{b}-\bar{c}), (\bar{c}-\bar{a})$.

$$Solution: P=0.$$

1.3.11. From the development in two different ways of the product $(\bar{a}\times\bar{b})(\bar{c}\times\bar{u})$ deduce the components of $\bar{u}$ along the directions of the vectors $\bar{a}, \bar{b}$ and $\bar{c}$, considered non-coplanar.

$$Solution: \bar{u}=\frac{1}{(\bar{a}\bar{b}\bar{c})}\left[(\bar{b}\bar{c}\bar{u})\bar{a}+(\bar{c}\bar{a}\bar{u})\bar{b}+(\bar{a}\bar{b}\bar{u})\bar{c}\right].$$

1.3.12. If the non-coplanar vectors $\bar{a}, \bar{b}$ and $\bar{c}$ are given, constructed with these vectors $\bar{u}$, $\bar{v}$ and $\bar{w}$ as follows:

$$\bar{u}=\bar{a}\times(\bar{b}+\bar{c}),\quad \bar{v}=\bar{b}\times(\bar{c}+\bar{a}),\quad \bar{w}=\bar{c}\times(\bar{a}+\bar{b}).$$

Show that:

(a) $\bar{u}\times\bar{v}=\bar{v}\times\bar{w}=\bar{w}\times\bar{u}=(\bar{a}\bar{b}\bar{c})(\bar{a}+\bar{b}+\bar{c})$.
(b) $P=(\bar{u}\bar{v}\bar{w})=0$.

1.3.13. Write the equation of a straight line for the following cases:

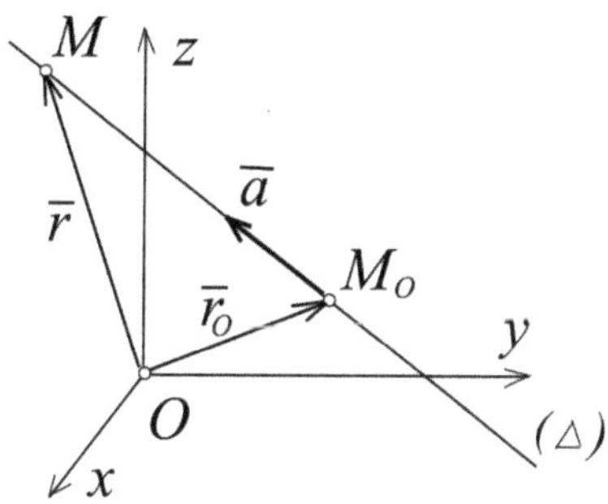

Fig. 1.18 Straight line passing through a point M_o

(a) It passes through the point $M_o(\bar{r}_o)$ and has the direction given by the vector $\bar{a}$;
(b) It passes through the points $A(\bar{r}_A)$ and $B(\bar{r}_B)$;
(c) It passes through the origin and has the direction given by the vector $\bar{a}$.

Solution: (a) M being a certain point on the line (Δ), we have (see Fig. 1.18): $\overrightarrow{M_oM} = \lambda\vec{a}$ with $\lambda \in \mathbf{R}$. The relation can be written:

$$\bar{r} - \bar{r}_o = \lambda\bar{a}, \lambda \in \mathbf{R},$$

or

$$\bar{r} = \bar{r}_o + \lambda\bar{a}, \lambda \in \mathbf{R}.$$

If the vectors $\bar{r}(x, y, z)$, $\bar{r}_o(x_o, y_o, z_o)$ and $\bar{a}\left(a_x, a_y, a_z\right)$ are defined by their Cartesian coordinates, the equation of the straight line takes the form:

$$x = x_o + \lambda a_x; \quad y = y_o + \lambda a_y; \quad z = z_o + \lambda a_z, \quad \lambda \in R$$

or, removing the parameter λ:

$$\frac{x - x_o}{a_x} = \frac{y - y_o}{a_y} = \frac{z - z_o}{a_z}.$$

Premultipling in a cross product, at right with $\bar{a}$ the vectorial relation $\bar{r} = \bar{r}_o + \lambda\bar{a}$, one obtains:

$$\bar{r} \times \bar{a} = \bar{r}_o \times a.$$

Note $\bar{b} = \bar{r}_o \times \bar{a}(\bar{b}$ is a vector perpendicular to $\bar{a}$ that $\bar{a} \cdot \bar{b} = 0)$. Then, the vector equation of the straight line can be written as:

$$\bar{r} \times \bar{a} = \bar{b} \quad \left(\text{with } \bar{a} \cdot \bar{b} = 0\right).$$

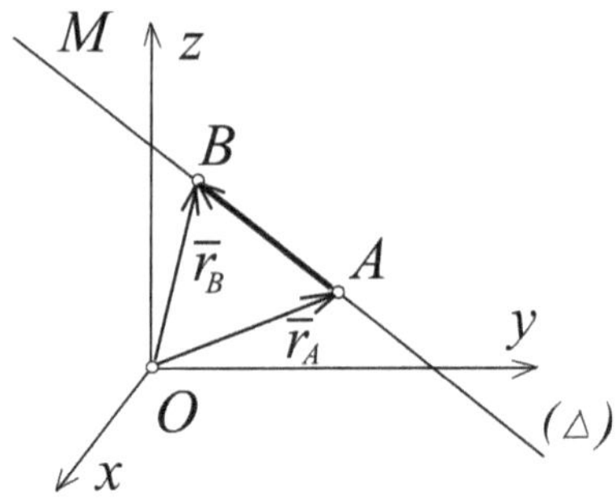

Fig. 1.19 Straight line through the points A and B

(b) The vector $\overrightarrow{AB} = \vec{r}_B - \vec{r}_A$ is collinear with the straight line (see Fig. 1.19), so if M is a point on the line, it can be written: $\overrightarrow{M_oM} = \lambda\overrightarrow{AB}, \lambda \in \boldsymbol{R}$, or $\bar{r} - \bar{r}_o = \lambda(\bar{r}_B - \bar{r}_A)$ or else:

$$\bar{r} = \bar{r}_o + \lambda(\bar{r}_B - \bar{r}_A).$$

If we consider the vectors $\bar{r}_A(x_A, y_A, z_A), \bar{r}_B(x_B, y_B, z_B)$ defined by their Cartesian coordinates, the equation of the straight line takes the form:

$$x = x_o + \lambda(x_B - x_A); \quad y = y_o + \lambda(y_B - y_A); \quad z = z_o + \lambda(z_B - z_A), \quad \lambda \in \mathbf{R}$$

or, if λ is removed:

$$\frac{x - x_o}{x_B - x_A} = \frac{y - y_o}{y_B - y_A} = \frac{z - z_o}{z_B - z_A}.$$

If we start now from the vectorial relationship multiplied to the right in a cross product with $\bar{r}_B - \bar{r}_A$, we get:

$$\bar{r} \times (\bar{r}_B - \bar{r}_A) = \bar{r}_A \times (\bar{r}_B - \bar{r}_A) = \bar{r}_A \times \bar{r}_B,$$

or

$$\bar{r} \times (\bar{r}_B - \bar{r}_A) = \bar{r}_A \times \bar{r}_B.$$

(c) Using the previously obtained result with $\bar{r}_A = 0$ (point A is the same with origin O) we have:

$$\bar{r} \times \bar{r}_B = 0.$$

In this case $\bar{r}_B$ indicates the direction of the line so $\bar{r}_B = \lambda\bar{a}$ and then:$\bar{r} \times \bar{a} = 0$.

1.3.14. Write the equation of a straight line parallel to the Ox axis.

$$\textit{Solution}: \ \bar{r} \times \bar{i} = \bar{b} \ \text{ with } \ \bar{b} = \alpha \bar{j} + \beta \bar{k}.$$

1.3.15. Write the condition that three points are collinear.

Solution: Let us consider the points A, B and C. The condition that they are collinear can be written as:

$$\overline{AC} = \mu \overline{AB},$$

or

$$\bar{r}_C - \bar{r}_A = \lambda(\bar{r}_B - \bar{r}_A).$$

If we premultiply to the left in a cross product with $\bar{r}_B - \bar{r}_A$, one obtains: $(\bar{r}_B - \bar{r}_A) \times (\bar{r}_C - \bar{r}_A) = 0$ or in a symmetrical form:

$$\bar{r}_A \times \bar{r}_B + \bar{r}_B \times \bar{r}_C + \bar{r}_C \times \bar{r}_A = 0.$$

1.3.16. (a) Write the equation of the plane π that passes through the points $A(\bar{r}_A), B(\bar{r}_B), C(\bar{r}_C)$. Numeric application: $\bar{r}_A(1, 3, 5), \bar{r}_B(3, 2, 1), \bar{r}_C(2, 1, 5)$;
(b) Write the equation of the plane π passing through point A and the normal direction is given by the vector $\bar{a}$;
(c) What is the condition for the points $A(\bar{r}_A), B(\bar{r}_B), C(\bar{r}_C), D(\bar{r}_D)$ to be coplanar?

Solution: We have: $\overline{AB} = \bar{r}_B - \bar{r}_A$; $\overline{AM} = \bar{r} - \bar{r}_A$. A vector perpendicular to the plane will be perpendicular to the vectors $\overline{AB}; \overline{AC}; \overline{AM}$ (Fig. 1.20). If $\bar{n}$ is normal to the plane.

$$\bar{n} = \overline{AB} \times \overline{AC} = \bar{r}_A \times \bar{r}_B + \bar{r}_B \times \bar{r}_C + \bar{r}_C \times \bar{r}_A,$$

then we must have $\overline{AM} \times \bar{n} = 0$ or:

$$(\bar{r} - \bar{r}_A)[(r_B - \bar{r}_A) \times (r_C - \bar{r}_A)] = 0.$$

After developments, we get:

$$\bar{r}[(\bar{r}_B - \bar{r}_A) \times (\bar{r}_C - \bar{r}_A)] = \bar{r}_A[(\bar{r}_B - \bar{r}_A) \times (\bar{r}_C - \bar{r}_A)],$$

or

$$\bar{r}(\bar{r}_A \times \bar{r}_B + \bar{r}_B \times \bar{r}_C + \bar{r}_C \times \bar{r}_A) = [\bar{r}_A \bar{r}_B \bar{r}_C],$$

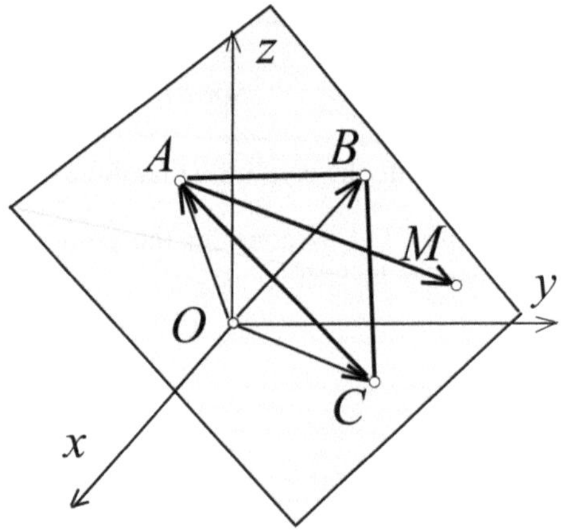

Fig. 1.20 A plane passing through three points A, B and C

or

$$\bar{r}n = [\bar{r}_A \bar{r}_B \bar{r}_C].$$

Numeric application:

$$\bar{n} = \begin{vmatrix} \bar{i} & \bar{j} & k \\ 3-1 & 2-3 & 1-5 \\ 2-1 & 1-3 & 5-5 \end{vmatrix} = -8\bar{i} - 4\bar{j} + 3\bar{k},$$

$$[\bar{r}_A \bar{r}_B \bar{r}_C] = \begin{vmatrix} 1 & 3 & 5 \\ 3 & 2 & 1 \\ 2 & 1 & 5 \end{vmatrix} = -35,$$

so the plane equation will be:

$$\bar{r}\left(-8\bar{i} - 4\bar{j} + 3\bar{k}\right) = -35,$$

or

$$8x + 4y - 3z = 35.$$

(b) If $\vec{n}$ it is the normal vector to the plane, it is perpendicular to any vector in plane, hence: $(\bar{r} - \bar{r}_A)\bar{n} = 0$ or $\bar{r}\bar{n} = \bar{r}_A\bar{n}$. In our case

$$r\bar{a} = \bar{r}_A\bar{a}.$$

(c) Let $\bar{n} = \bar{r}_A \times \bar{r}_B + \bar{r}_B \times \bar{r}_C + \bar{r}_C \times \bar{r}_A$ be a normal vector to the plane defined by A, B, C (see point a). If D there is in plan, we must have:

$$(\bar{r}_D - \bar{r}_A)\bar{n} = 0,$$

or

$$\bar{r}_D\bar{n} = [\bar{r}_A\bar{r}_B\bar{r}_C].$$

We have:

$$\bar{r}_D[\bar{r}_A \times \bar{r}_B + \bar{r}_B \times \bar{r}_C + \bar{r}_C \times \bar{r}_A] = [\bar{r}_A\bar{r}_B\bar{r}_C],$$

or

$$[r_A\bar{r}_B\bar{r}_C] + [\bar{r}_B\bar{r}_C\bar{r}_D] + [\bar{r}_C\bar{r}_D\bar{r}_A] + [\bar{r}_D\bar{r}_A\bar{r}_B] = 0.$$

1.3.17. Solve the vector equation $\bar{r} \times a = \bar{b}$.

Solution: If we pre-multiply the vector equation in a cross product with $\vec{a}$ it is obtained:

$$\bar{a} \times (\bar{r} \times \bar{a}) = \bar{a} \times \bar{b}.$$

or, expanding the vector triple product:

$$a^2\bar{r} - (ar)\bar{a} = \bar{a} \times \bar{b},$$

from where:

$$\bar{r} = \frac{\bar{a} \times \bar{b}}{a^2} + \lambda\bar{a}, \quad \text{with} \quad \lambda = \frac{(\bar{a}\bar{r})}{a^2}.$$

1.3.18. Let us show that $d = \left|\frac{\bar{a}\times\bar{b}}{a^2}\right| = \frac{b}{a}$ represents the distance from origin of an Euclidian system to the line $\bar{r} \times \bar{a} = \bar{b}(\bar{a}\bar{b} = 0)$.

Solution: The equation $\bar{r} \times \bar{a} = \bar{b}$ has the solution: $\bar{r} = \frac{\bar{a}\times\bar{b}}{a^2} + \lambda\bar{a}$. If $\lambda = 0$ the vector:

$$r|_{\lambda=0} = \bar{d} = \frac{\bar{a} \times \bar{b}}{a^2},$$

connects the origin of the coordinate system with a point on the line and is perpendicular to the line, since it represents the vector whose magnitude is the distance from the origin to the line.

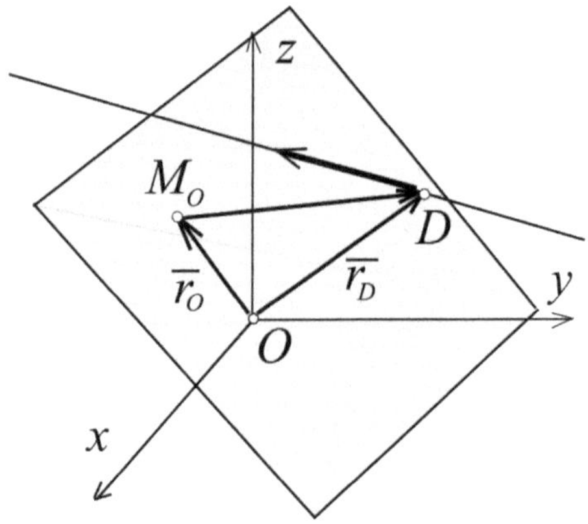

Fig. 1.21 A plane containing a line and a point M_o

1.3.19. Determine the equation of the plane that passes through the origin of the coordinate system and is perpendicular to the vector $\bar{a}$.

Solution: In problem 1.3.16(b) one gets $\bar{r}_A = 0$ and $\bar{r}\bar{a} = 0$. If the coordinate planes yOz, zOx, xOy, are considered they have the equations: $\bar{r}\bar{i} = 0$; $\bar{r}\bar{j} = 0$; $\bar{r}\bar{k} = 0$.

1.3.20. Write the vectorial form of the equation of a plane that intersects the axes in the points $A(\frac{1}{a}, 0, 0)$, $B(0, \frac{1}{b}, 0)$, $C(0, 0, \frac{1}{c})$.

Solution: The plan equation is in this case:

$$xa + yb + zc = 1.$$

If the vectors $\bar{r} = x\bar{i} + y\bar{j} + z\bar{k}$ and $\bar{n} = a\bar{i} + b\bar{j} + c\bar{k}$ are considered, it can be written: $\bar{r}\bar{n} = 1$.

1.3.21. Write the equation of the plane containing the line $\bar{r} \times \bar{a} = \bar{b}$, $(\bar{a}\bar{b} = 0)$ and the point $\bar{r}_o$ (see Fig. 1.21).

Solution: By solving the equation of the line we obtain:

$$\bar{r} = \frac{\bar{a} x \bar{b}}{a^2} + \lambda \bar{a}.$$

Point D for which $\lambda = 0$ belongs to the following plane:

$$\bar{r}_D = \frac{\bar{a} \times \bar{b}}{a^2}.$$

A normal vector to the plane will be given by:

$$\bar{n} = (\bar{r}_D - \bar{r}_o)x\bar{a} = \bar{r}_D = \left(\frac{\bar{a}x\bar{b}}{a^2} - \bar{r}_o\right)x\bar{a} = \bar{b} - \bar{r}_o x\bar{a}.$$

If we write the plan equation in the form (see 1.1.15b)

$$\bar{r}\bar{n} = \bar{r}_o\bar{n},$$

and introducing $\bar{n}$, one obtains:

$$\bar{r}\left(\bar{b} - \bar{r}_o \times \bar{a}\right) = \left(\bar{b}\bar{r}_o\right).$$

1.3.22. Write the equation of a plane perpendicular to the Oz axis and passing through the point $\bar{r}_A$.

$$Solution\text{: } \bar{r}\bar{k} = \bar{r}_A\bar{k} = z_A.$$

1.3.23. Determine the point of intersection between the plane $\bar{r} \cdot \bar{a}_1 = c$ and the line $\bar{r} \times \bar{a}_2 = \bar{b}$, $\left(\bar{a}_2\bar{b} = 0\right)$.

Application:

$$\bar{a}_1(1, 3, 1),\ \bar{a}_2(2, 1, -1),\ \bar{b}(1, -1, 1),\ c = 1.$$

Solution: We pre-multiply, to the left, in a cross product with $\bar{a}_1$, the line equation. It is obtained:

$$\bar{a}_1 \times (\bar{r} \times a_2) = \bar{a}_1 \times \bar{b}.$$

Considering the vector triple product expansion, one obtains:

$$(\bar{a}_1\bar{a}_2)\bar{r} - (\bar{a}_1\bar{r})\bar{a}_2 = \bar{a}_1 \times \bar{b},$$

from where:

$$\bar{r} = \frac{\bar{a}_1 \times \bar{b} + c\bar{a}_2}{(\bar{a}_1\bar{a}_2)}.$$

For the given values:

$$\bar{a}_1 \times \bar{b} = 4\bar{i} - 4\bar{k};\quad (\bar{a}_1 a_2) = 4;\quad \bar{r}_I = \frac{3}{2}\bar{i} + \frac{1}{4}\bar{j} - \frac{5}{4}\bar{k}.$$

1.3.24. Determine the points of intersection of the straight line $\bar{r} \times \bar{a} = \bar{b}$, $(\bar{a}\bar{b} = 0)$ with the coordinate planes (x=0, y=0, z=0). Application: $\bar{a}(1, 3, -2)$; $\bar{b}(2, 2, 4)$

Solution: For the intersection with the yOz $\left(\bar{r}\cdot\bar{i}=0\right)$ plane, the equation of the straight is multiplied by $\vec{i}$ in a dot product. It is obtained:

$$\bar{i}\times(\bar{r}\times\bar{a})=i\times\bar{b}.$$

Expanding the vector triple product, one obtains:

$$\left(\bar{a}\bar{i}\right)\bar{r}-\left(\bar{r}\bar{i}\right)\bar{a}=\bar{i}\times\bar{b},$$

$$\bar{r}_{xOy}=\frac{\bar{i}\times\bar{b}}{\left(\bar{a}\bar{i}\right)}+\frac{\left(\bar{r}\bar{i}\right)}{\left(\bar{a}\bar{i}\right)}\bar{a}=\frac{\bar{i}\times\bar{b}}{\left(\bar{a}\bar{i}\right)}.$$

Analogously, the other two intersections are obtained. For given values:

$$\begin{aligned}&\bar{i}\times\bar{b}=-4\bar{j}+2k;\quad \bar{j}\times\bar{b}=4\bar{i}-2\bar{k};\quad \bar{k}\times\bar{b}=-2\bar{i}+2;\\&\left(\bar{a}\bar{i}\right)=1;\quad \left(\bar{a}\,\bar{j}\right)=3;\quad \left(\bar{a}\bar{k}\right)=-2.\end{aligned}$$

It results:

$$\bar{r}_{xOy}=-4\bar{j}+2\bar{k};\quad \bar{r}_{yOz}=\frac{4}{3}\bar{i}-\frac{2}{3}\bar{k};\quad \bar{r}_{zOx}=\bar{i}-\bar{k}.$$

1.3.25. Determine the line intersection of planes: $\bar{r}\cdot a_1=c_1$; $\bar{r}\cdot\bar{a}_2=c_2$ (see Fig. 1.22).

Solution: If we write the line as: $\bar{r}\times\bar{a}=\bar{b}$, $(\bar{a}\bar{b}=0)$, the vector $\bar{a}$ will belong to the two planes. If $\bar{a}_1$ and $\bar{a}_2$ are perpendicular to the two planes, one can write: $\bar{a}=\bar{a}_1\times\bar{a}_2$. It remains $\bar{b}$ to be determined. The equation of the line is, if $\bar{a}$ is known: $\bar{r}\times(\bar{a}_1\times\bar{a}_2)=\bar{b}$. By developing the vector triple product, it results:

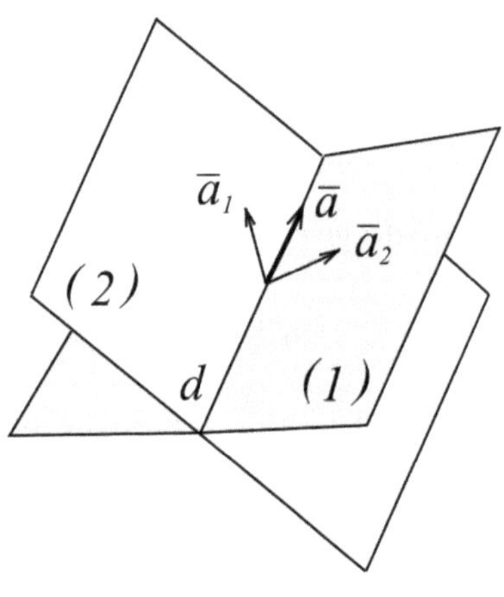

Fig. 1.22 Intersection of two plane is a straight line

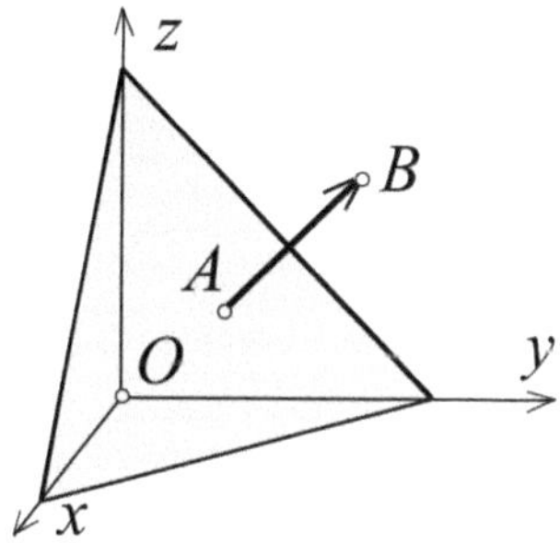

Fig. 1.23 Distance from a point to a plane

$$(\bar{r}\bar{a}_2)\bar{a}_1 - (\bar{r}\bar{a}_1)\bar{a}_2 = \bar{b}.$$

Taking into account the relations of the planes definition, we obtain:

$$\bar{b} = c_2\bar{a}_1 - c_1\bar{a}_2.$$

The equation of the intersection line follows in the vector triple product form:

$$\bar{r} \times (\bar{a}_1 \times \bar{a}_2) = c_2\bar{a}_1 - c_1\bar{a}_2.$$

1.3.26. Let us consider: $\bar{r}\cdot\bar{a}_1 = c_1, \bar{r}\cdot\bar{a}_2 = c_2, \bar{r}\cdot\bar{a}_3 = c_3$. Determine the intersection of the planes $\bar{r}\cdot\bar{a}_1 = c_1, \bar{r}\cdot\bar{a}_2 = c_2, \bar{r}\cdot\bar{a}_3 = c_3$.

Solution: According to the previous problem (1.3.25), the intersection line of the first two planes is:

$$\bar{r} \times (\bar{a}_1 \times \bar{a}_2) = c_2\bar{a}_1 - c_1\bar{a}_2$$

which must be intersected with the plan:

$$\bar{r}\cdot\bar{a}_3 = c_3.$$

We pre-multiply, in a cross product, the line equation with $\bar{a}_3$. Considering the expansion of the vector triple product, it is obtained:

$$[\bar{a}_3(\bar{a}_1 \times \bar{a}_2)]\bar{r} - (\bar{a}_3\bar{r})(\bar{a}_1 \times \bar{a}_2) = c_2(\bar{a}_3 \times \bar{a}_1) - c_1(\bar{a}_3 \times \bar{a}_2),$$

from where the point of intersection results:

$$\bar{r}_I = \frac{c_1(\bar{a}_2 \times \bar{a}_3) + c_2(\bar{a}_3 \times \bar{a}_1) + c_3(\bar{a}_1 \times \bar{a}_2)}{[\bar{a}_1\bar{a}_2\bar{a}_3]}.$$

1.3.27. Determine the distance from a point $B(\bar{r}_B)$ to the plane $\bar{r}\cdot\bar{a} = c$ (see Fig. 1.23).

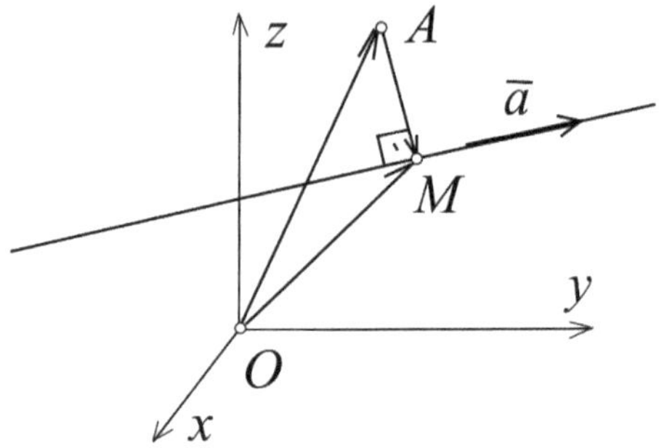

Fig. 1.24 Distance from a point A to a line

Solution: Let A be the foot of the perpendicular from B to the plane. We have: $\overline{AB} = \lambda\bar{a}$ or $\bar{r}_B - \bar{r}_A = \lambda\bar{a}$. If this relationship is multiplied in a dot product with $\bar{a}$, we obtain:

$$(\bar{r}_B - \bar{r}_A)\bar{a} = \lambda a^2.$$

Hence:

$$\lambda = \frac{\bar{r}_B\bar{a} - \bar{r}_A\bar{a}}{a^2} = \frac{\bar{r}_B\bar{a} - c}{a^2}.$$

The distance is:

$$d = \left|\overrightarrow{AB}\right| = \lambda a = \frac{\bar{r}_B\bar{a} - c}{a}.$$

1.3.28. Determine the distance from point $A(\bar{r}_A)$ to the line $\bar{r} \times \bar{a} = \bar{b}$ (see Fig. 1.24).

Solution: Let $M(\bar{r})$ the perpendicular foot from A to the line. The vector is perpendicular to the unit vector of the line. Thus, it can be written:

$$\overline{AM} \cdot \bar{a} = 0,$$

or

$$(\bar{r} - \bar{r}_A) \cdot \bar{a} = 0$$
$$\bar{r} \cdot \bar{a} = \bar{r}_A \cdot \bar{a}.$$

If the vectorial form of the equation of the line is pre-multiplied by vector $\bar{a}$, it is obtained, after the vector triple product expansion:

$$a^2\bar{r} - (\bar{a}\bar{r})\bar{a} = \bar{a} \times \bar{b},$$

or

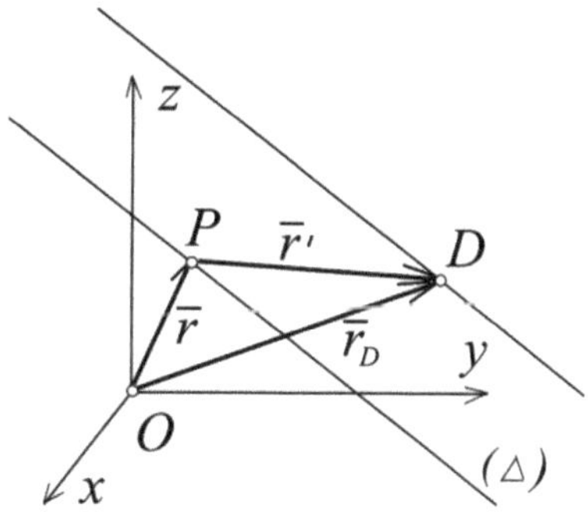

Fig. 1.25 The points on which the moment is known $(\bar{M})$ are situated on a straight line

$$\bar{r} = \frac{\bar{a} \times \bar{b}}{a^2} + \frac{(\bar{a}\bar{r})}{a^2}\bar{a} = \frac{\bar{a} \times \bar{b}}{a^2} + \frac{(\bar{a}\bar{r}_A)}{a^2}\bar{a} = \bar{r}_M,$$

which represents the position vector of point M. Then the distance vector between point A and line is:

$$\bar{\delta} = \overline{AM} = \bar{r} - \bar{r}_A = \frac{\bar{a} \times \bar{b}}{a^2} + \frac{(\bar{a}\bar{r}_A)}{a^2}\bar{a} - \bar{r}_A,$$

and its magnitude is:

$$\begin{aligned}\delta &= \frac{\left|\bar{a} \times \bar{b} + (\bar{a}\bar{r}_A)\bar{a} - a^2\bar{r}_A\right|}{a^2} \\ &= \frac{\sqrt{\sum \left[a_y b_z - a_z b_y + a_x(x_A a_x + y_A a_y + z_A a_z) - (a_x^2 + a_y^2 + a_z^2)x_A\right]^2}}{a^2}.\end{aligned}$$

1.3.29. Determine the distance between the lines $\bar{r} \times \bar{a}_1 = \bar{b}_1$ and $\bar{r} \times \bar{a}_2 = \bar{b}_2$ (see Fig. 1.25).

Solution: Let $\overline{AB}$ be the distance vector between the two lines. Then, we will have the relationships: $\overline{AB} \cdot \bar{a}_1 = 0$ and $\overline{AB} \cdot \bar{a}_2 = 0$. It follows that we must have the relationship: $\overline{AB} = \bar{r}_B - \bar{r}_A = \lambda(\bar{a}_1 \times \bar{a}_2)$ and $\bar{r}_A \times \bar{a}_1 = \bar{b}_1, \bar{r}_B \times \bar{a}_2 = \bar{b}_2$. If we scalar multiply the vector $\overrightarrow{AB}$ we get:

$$\overline{AB} \cdot (\bar{a}_1 \times \bar{a}_2) = \lambda(\bar{a}_1 \times \bar{a}_2)^2,$$

or

$$\begin{aligned}(\bar{r}_B - \bar{r}_A)(\bar{a}_1 \times \bar{a}_2) &= \lambda(\bar{a}_1 \times \bar{a}_2)^2 \\ \bar{r}_B(\bar{a}_1 \times \bar{a}_2) - \bar{r}_A(\bar{a}_1 \times \bar{a}_2) &= \lambda(\bar{a}_1 \times \bar{a}_2)^2.\end{aligned}$$

If the properties of the scalar triple product are taken into account, one can read:

$$-\bar{a}_1(\bar{r}_B \times \bar{a}_2) - \bar{a}_2(\bar{r}_A \times \bar{a}_1) = \lambda(\bar{a}_1 \times \bar{a}_2)^2.$$

By introducing the equations of the lines in brackets, it follows:

$$-\bar{a}_1\bar{b}_2 - \bar{a}_2\bar{b}_1 = \lambda(\bar{a}_1 \times \bar{a}_2)^2,$$

from where we deduce:

$$\lambda = \frac{-\bar{a}_1\bar{b}_2 - \bar{a}_2\bar{b}_1}{(\bar{a}_1 \times \bar{a}_2)^2}.$$

The distance between the two straight lines is thus obtained:

$$\overline{AB} = \frac{-\bar{a}_1\bar{b}_2 - \bar{a}_2\bar{b}_1}{(\bar{a}_1 \times \bar{a}_2)^2}(\bar{a}_1 \times \bar{a}_2),$$

and its magnitude:

$$AB = \frac{\left|-\bar{a}_1\bar{b}_2 - \bar{a}_2\bar{b}_1\right|}{|(\bar{a}_1 \times \bar{a}_2)|}.$$

1.3.30. Determine the angle between the line $\bar{r}x\bar{a} = \bar{b}, (\bar{a}\bar{b} = 0)$ and the plane $\vec{r} \cdot \vec{m} = c$.

Solution: Let $A' \in \pi$ be the perpendicular foot from A, belonging to the line, to the plane. We have $AA' \perp \pi$, then $\overrightarrow{AA'} = \lambda\bar{m}$. It results $\bar{r}_{A'} - \bar{r}_A = \lambda\bar{m}$ or $\bar{r}_{A'} = \bar{r}_A + \lambda\bar{m}$. If we pre-multiply the equation of the straight line with $\bar{m}$, it results: $\bar{m}x(\bar{r}x\bar{a}) = \bar{m}x\bar{b}$. If the vector triple product develops, there is obtained:

$$\bar{r} = \frac{\bar{m} \times b - (\bar{m}\bar{r})\bar{a}}{(\bar{m}\bar{a})},$$

and if we impose the condition that the intersection point belongs to the plan $(\bar{r} \cdot \bar{m} = c)$, one obtains the position vector of the point of intersection of the straight line with the plane:

$$\bar{r}_I = \frac{\bar{m}x\bar{b} - c\bar{a}}{(\bar{m}\bar{a})}.$$

Let us find the direction of the straight line $A'I$. We have:

$$\bar{a}_1 = \bar{r}_I - \bar{r}_A = \frac{\bar{m} \times \bar{b} - c\bar{a}}{(\bar{m}\bar{a})} - \frac{\bar{a} \times \bar{b}}{a^2}.$$

From the vectorial equation of the line:

$$\bar{r} = \frac{\bar{a} \times \bar{b}}{a^2} + \lambda_1 \bar{a},$$

it is considered for the choice of point A the particular case when $\lambda_1 = 0$, therefore:

$$\bar{r}_A = \frac{\bar{a} \times \bar{b}}{a^2}.$$

We have

$$(\bar{a}\bar{a}_1) = a a_1 \cos(< \bar{a}\bar{a}_1),$$

from where

$$\cos(< \bar{a}\bar{a}_1) = \frac{(\bar{a}\bar{a}_1)}{a a_1}.$$

But

$$(\bar{a}\bar{a}_1) = \frac{\bar{a}\left(\bar{m} \times \bar{b}\right) - ca^2}{(\bar{m}\bar{a})} - \frac{\bar{a}\left(\bar{a} \times \bar{b}\right)}{a^2} = \frac{\bar{m}\left(\bar{b} \times \bar{a}\right) - ca^2}{(\bar{m}\bar{a})}.$$

In this case, it follows:

$$\cos(< \bar{a}\bar{a}_1) = \frac{\frac{\bar{m}\left(\bar{b}\times\bar{a}\right) - ca^2}{(\bar{m}\bar{a})}}{a\left|\frac{\bar{m}\times\bar{b} - c\bar{a}}{(\bar{m}\bar{a})} - \frac{\bar{a}\times\bar{b}}{a^2}\right|}.$$

1.3.31. Given the line on which the sliding force is $\bar{F}$, determine the points in which the moment vector is $\bar{M}\left(\bar{F} \cdot \bar{M} = 0\right)$, (see Fig. 1.25).

Solution: Let P be a point on the force support (where the moment is null) and D a point where the moment vector is $\bar{M}$. We have:

$$\bar{M} = \bar{F} \times \bar{r}'.$$

If we pre-multiply vectorally with $\bar{F}$ we get:

$$\bar{r}' = -\frac{\bar{F} \times \bar{M}}{F^2} + \lambda' \bar{F}.$$

The position vector of the point D is:

$$\bar{r}_D = \bar{r} + \bar{r}' = \frac{\bar{a} \times \bar{b}}{a^2} + \lambda \bar{a} - \frac{\bar{F} \times \bar{M}}{F^2} + \lambda' \bar{F}.$$

If it is taken into account that the force is colinear with the vector $\bar{a}\left(\bar{F} = \gamma \cdot \bar{a}\right)$, it results:

$$\bar{r}_D = \frac{\bar{a} \times \left(\bar{b} - \frac{\gamma \bar{M}}{\gamma^2}\right)}{a^2} + (\lambda + \lambda' \gamma)\bar{a} = \frac{\bar{a} \times \bar{c}}{a^2} + \lambda'' \bar{a},$$

where: $\bar{c} = \bar{b} - \frac{\gamma \bar{M}}{\gamma^2}$ and $\lambda'' = \lambda + \lambda' \gamma$ that is, a parallel line with the force.

References

Liesen, J., Mehrmann, V.: Linear Algebra. Springer (2015)
Simionescu, G.D.: Vector Algebra and Applications in Geometry. Tehnica, Bucharest (1982)
Springer, C.E.: Tensor and Vector Analysis: with Applications to Differential Geometry. Dover Publication Inc., Mineola, New York (2013)

Chapter 2
Matrices

2.1 Fundamental Notions

We introduce in the following the basic notions about matrices and the basic properties interesting for our presentation. A number of other properties will be presented in the following chapters. A matrix of dimension $m \times n$ is represented by a table with m lines and n columns (in box bracket or parentheses):

$$\begin{bmatrix} a_{11} & a_{12} & \dots & a_{1n} \\ a_{21} & a_{22} & \dots & a_{2n} \\ \vdots & & & \\ a_{m1} & a_{m2} & \dots & a_{mn} \end{bmatrix}, \begin{Vmatrix} a_{11} & a_{12} & \dots & a_{1n} \\ a_{21} & a_{22} & \dots & a_{2n} \\ \vdots & & & \\ a_{m1} & a_{m2} & \dots & a_{mn} \end{Vmatrix}, \begin{pmatrix} a_{11} & a_{12} & \dots & a_{1n} \\ a_{21} & a_{22} & \dots & a_{2n} \\ \vdots & & & \\ a_{m1} & a_{m2} & \dots & a_{mn} \end{pmatrix}.$$

There are three different ways found in literature to write these sizes, but in this book we will prefer the notation with box brackets as it is easy to use.

The numbers of rows and columns of a matrix define its size. If a matrix has m rows and n columns, it is called a m-by-n matrix (or $m \times n$ matrix). The number m and n are called the dimensions of the matrix.

Notation of Matrices

There are different notations for matrices found in literature. Some notations which have become more used are outlined below. The set of matrices with m lines and n columns is denoted by $\mathcal{M}_{m \times n}$. Uppercase letters, such as A …, are used to denote matrices. These uppercase letters are sometimes written in bold as: $\mathbf{A}_{m \times n}$, $\mathbf{B}_{m \times n}$, $\mathbf{K}_{m \times n}$. It is also customary to insert the symbol between parentheses: $[\mathrm{A}]_{m \times n}$, $[\mathrm{B}]_{m \times n}$, $[\mathrm{K}]_{m \times n}$ or representations of the form: $[a_{ij}]$, $\|a_{ij}\|$, (a_{ij}). For a matrix having a single column, the notation with braces is used (these matrices represent vectors): $\{A\}, \{B\}, \{X\}$.

S. Vlase et al., *Eigenvalue and Eigenvector Problems in Applied Mechanics*, Advanced Structured Materials 96, https://doi.org/10.1007/978-3-030-00991-5_2

A matrix with a single row is called a row vector and a matrix with a single column is called a column vector.

Square matrix: A matrix at which the number of rows is equal to the number of columns is called a square matrix. If the number of rows and columns is equal to n, we have a matrix of size n. For example:

$$[A] = \begin{bmatrix} 1 & 2 & 3 & 4 \\ -1 & 2 & 2 & 2 \\ 3 & 5 & 1 & -1 \\ 0 & 3 & -1 & 2 \end{bmatrix}.$$

The set of squares matrix of dimension n is denoted by $\mathcal{M}_n$. A square matrix is denoted by A_n or, if the dimension is understood, it is simply written, A. To define a square matrix, it is necessary to know its n^2 elements.

Symmetric matrix. A matrix whose elements have the property $a_{ij} = a_{ji}$ is called a symmetric matrix, for example:

$$[A] = \begin{bmatrix} 1 & 2 & 3 & 4 \\ 2 & 0 & -2 & 5 \\ 3 & -2 & 1 & -1 \\ 4 & 5 & -1 & 2 \end{bmatrix}.$$

Observe that the elements of the matrix are symmetrical with respect to the first diagonal. For the definition of a symmetric matrix it is necessary to know only the elements on the diagonal and above it, so only $1 + 2 + 3 + \cdots + n = \frac{n(n+1)}{2}$ elements versus n^2 are needed in the case of any matrix.

Skew-symmetric matrix. A matrix whose elements have the property $a_{ij} = -a_{ji}$ is called a skew-symmetric (antisymmetric or antimetric) matrix. If $i = j$ then the definition relation becomes: $a_{ii} = -a_{ii}$ or $2a_{ii} = 0$ so the elements on the main diagonal of an antisymmetric matrix are zero, for example:

$$[A] = \begin{bmatrix} 0 & 2 & 3 & 4 \\ -2 & 0 & -2 & 5 \\ -3 & 2 & 0 & -1 \\ -4 & -5 & 1 & 0 \end{bmatrix},$$

To define a skew-symmetric matrix, it is necessary to know only the elements above or below the diagonal, so $\frac{n(n-1)}{2}$ elements.

Diagonal matrix. A matrix with non-zero elements only on the main diagonal is called a diagonal matrix. The relation of definition of a diagonal matrix is: $a_{ij} = 0$ if $i \neq j$ and $a_{ij} \neq 0$ if $i = j$.

For example:

$$[A] = \begin{bmatrix} 1 & 0 & 0 & 0 \\ 0 & 2 & 0 & 0 \\ 0 & 0 & 3 & 0 \\ 0 & 0 & 0 & 5 \end{bmatrix}.$$

To define a diagonal matrix, we only need to know the elements on the diagonal, so n elements.

Triangular matrix. A matrix that has nonzero elements on the diagonal and above or below is called a triangular matrix. A matrix is an upper triangular matrix if the elements below the main diagonal are zero:

$$[T_S] = \begin{bmatrix} 1 & 4 & 2 & -5 \\ 0 & 2 & 3 & 2 \\ 0 & 0 & 3 & -1 \\ 0 & 0 & 0 & 5 \end{bmatrix},$$

and a lower triangular matrix if the elements above the main diagonal are zero:

$$[T_i] = \begin{bmatrix} 1 & 0 & 0 & 0 \\ 4 & 2 & 0 & 0 \\ 2 & 3 & 3 & 0 \\ -5 & 2 & -1 & 5 \end{bmatrix}.$$

2.2 Basic Operation

2.2.1 Addition ($+\mathcal{M}_{m \times n} \times \mathcal{M}_{m \times n} \rightarrow \mathcal{M}_{m \times n}$)

Consider two matrices $[A], [B] \in \mathcal{M}_{m \times n}$. The sum of these two matrices, denoted by +, is a matrix $[C] \in \mathcal{M}_{m \times n}$ that reads:

$$[C] = [A] + [B], \tag{2.1}$$

where matrix $[C]$ elements are defined by the relationships: $c_{ij} = a_{ij} + b_{ij}$. For example, if:

$$[A] = \begin{bmatrix} 1 & 2 & 0 \\ -3 & 0 & -1 \end{bmatrix}; \ [B] = \begin{bmatrix} 2 & 2 & -2 \\ 1 & 3 & 4 \end{bmatrix},$$

one obtains:

$$[C] = [A] + [B] = \begin{bmatrix} 1 & 2 & 0 \\ -3 & 0 & -1 \end{bmatrix} + \begin{bmatrix} 2 & 2 & -2 \\ 1 & 3 & 4 \end{bmatrix}$$
$$= \begin{bmatrix} 1+2 & 2+2 & 0-2 \\ -3+1 & 0+3 & -1+4 \end{bmatrix} = \begin{bmatrix} 3 & 4 & -2 \\ -2 & 3 & -3 \end{bmatrix}.$$

For the sum there is the null element (additive identity), which we denote by $[O] \in \mathcal{M}_{m\times n}$ which, if $[A] \in \mathcal{M}_{m\times n}$, has the property:

$$[A] + [O] = [O] + [A] = [A]. \tag{2.2}$$

All the entries in the null matrix are zero:

$$[O]_{m\times n} = \begin{bmatrix} 0 & 0 & \dots & 0 \\ 0 & 0 & \dots & 0 \\ \vdots & & & \\ 0 & 0 & \dots & 0 \end{bmatrix}. \tag{2.3}$$

If we look at the zero matrix as a linear transformation, all the vectors after transformation are assigned to the zero vector. The null matrix is idempotent (when it is multiplied by itself the result is itself): The rank of a zero matrix is 0.

The addition is commutative, that is:

$$[A] + [B] = [B] + [A], \tag{2.4}$$

and associative:

$$([A] + [B]) + [C] = [A] + ([B] + [C]). \tag{2.5}$$

2.2.2 Scalar Multiplication ($\cdot R \times \mathcal{M}_{m\times n} \rightarrow \mathcal{M}_{m\times n}$)

Consider a real number, $\lambda \in R$, and a matrix $[A] \in \mathcal{M}_{m\times n}$. The scalar multiplication of the matrix $[A]$ with a scalar λ is a matrix $[C]$, which reads:

$$[C] = \lambda \cdot [A], \tag{2.6}$$

where the elements of $[C]$ are defined by the relationships: $c_{ij} = \lambda a_{ij}$. The scalar multiplication can be extended if λ belongs to the space of complex numbers. For example, if $\lambda = 2$ and:

$$[A] = \begin{bmatrix} 1 & 2 & 0 \\ -3 & 0 & -1 \end{bmatrix},$$

then the scalar multiplication by 2 gets:

$$[C] = \lambda \cdot [A] = 2 \begin{bmatrix} 1 & 2 & 0 \\ -3 & 0 & -1 \end{bmatrix} = \begin{bmatrix} 2 & 4 & 0 \\ -6 & 0 & -2 \end{bmatrix}.$$

Scalar multiplication is distributive in regards to the addition:

$$\lambda \cdot ([A] + [B]) = \lambda \cdot [A] + \lambda \cdot [B], \tag{2.7}$$

also:

$$\lambda \cdot (\mu \cdot [A]) = (\lambda \cdot \mu) \cdot [A],$$

and:

$$(\lambda + \mu) \cdot [A] = \lambda \cdot [A] + \mu \cdot [A].$$

2.2.3 Matrix Multiplication ($\cdot \mathcal{M}_{m\times p} \times \mathcal{M}_{p\times n} \rightarrow \mathcal{M}_{m\times n}$)

Consider $[A] \in \mathcal{M}_{m\times p}$ and $[B] \in \mathcal{M}_{p\times n}$. The product of these two matrices is a matrix $[C] \in \mathcal{M}_{m\times n}$ which we write in the form:

$$[C] = [A] \cdot [B], \tag{2.8}$$

where the entries of $[C]$ are defined by the relationships:

$$c_{ij} = \sum_k a_{ik} b_{kj}.$$

In this definition we emphasize the requirement that *the number of columns of the first matrix must be equal to the number of rows of the second*. If this is not met, the matrix product cannot be performed. This product is of the type of line on column. We can also define a column-type line product, but in the following we will work exclusively with this definition for the product of two matrices. For example, if:

$$[A]_{2\times 3} = \begin{bmatrix} 1 & 2 & 0 \\ -3 & 0 & -1 \end{bmatrix}; \quad [B]_{3\times 2} = \begin{bmatrix} 2 & 1 \\ 2 & 3 \\ -2 & 4 \end{bmatrix},$$

the product of the two matrices is a matrix of size 2×2 given by:

$$[C]_{2\times 2} = [A]\cdot[B] = \begin{bmatrix} 1 & 2 & 0 \\ -3 & 0 & -1 \end{bmatrix} \begin{bmatrix} 2 & 1 \\ 2 & 3 \\ -2 & 4 \end{bmatrix}$$

$$= \begin{bmatrix} 1\cdot 2+2\cdot 2+0\cdot(-2) & 1\cdot 2+2\cdot 3+0\cdot 4 \\ -3\cdot 2+0\cdot 2+(-1)\cdot(-2) & -3\cdot 1+0\cdot 3+(-1)\cdot 4 \end{bmatrix} = \begin{bmatrix} 6 & 8 \\ -4 & -7 \end{bmatrix}.$$

Generally, there is no neutral element (or identity element) for the multiplication of two certain matrices respecting the condition of existence of multiplication. Only for a square matrix we can find an identity matrix. If we consider the set of matrices of dimension n, then there is the neutral element $[E]$ for the multiplication operation, defined as follows:

$$[A]\cdot[E] = [E]\cdot[A] = [A]. \tag{2.9}$$

It obtains the identity matrix:

$$[E] = \begin{bmatrix} 1 & & & & \\ & 1 & & 0 & \\ & & \ddots & & \\ & 0 & & \ddots & \\ & & & & 1 \end{bmatrix}. \tag{2.10}$$

One cannot question the commutativity of two matrices of any size (which are not squares). If the matrices are square, in the general case, the multiplication is not commutative:

$$[A]\cdot[B] \neq [B]\cdot[A]. \tag{2.11}$$

A counter-example justifies the assertion.

Consider the following specific matrices

$$[A] = \begin{bmatrix} 1 & 2 \\ 3 & 4 \end{bmatrix} \text{ and } [B] = \begin{bmatrix} 5 & 6 \\ 7 & 8 \end{bmatrix}.$$

Then:

$$[A][B] = \begin{bmatrix} 1 & 2 \\ 3 & 4 \end{bmatrix} \begin{bmatrix} 5 & 6 \\ 7 & 8 \end{bmatrix} = \begin{bmatrix} 19 & 22 \\ 43 & 50 \end{bmatrix};$$

$$[B][A] = \begin{bmatrix} 5 & 6 \\ 7 & 8 \end{bmatrix} \begin{bmatrix} 1 & 2 \\ 3 & 4 \end{bmatrix} = \begin{bmatrix} 23 & 34 \\ 31 & 46 \end{bmatrix}.$$

The commutative matrices represent special cases and will not be encountered in the following.

The following properties occur for square matrices. Demonstration of these properties is immediate and is omitted here.

- The multiplication of matrices is distributive over their addition, i.e.:

$$([A] + [B])([C] + [D]) = [A][C] + [A][D] + [B][C] + [B][D].$$

- The product of two diagonal matrices is also a diagonal matrix;
- The product of two triangular matrices (upper or lower) is also a triangular matrix (upper or lower);
- We have the relationship:

$$[A]^{m+n} = [A]^m [A]^n.$$

- If $[O]$ is the zero matrix for the addition, we have:

$$[A][O] = [O].$$

- The square matrix multiplication is associative:

$$([A][B])[C] = [A]([B][C]).$$

Applications:

1. Show that $[A][B] = \det[A] \det[B]$.
2. If the matrix $[R]$ is of the form:

$$[R_\theta] = \begin{bmatrix} \cos\theta & -\sin\theta \\ \sin\theta & \cos\theta \end{bmatrix},$$

   demonstrate that:

$$[R_{\theta 1+\theta 2}] = [R_{\theta 1}][R_{\theta 2}].$$

3. If $[A][B] = [B][A]$ then we have:

$$[A]^m [B]^n = [B]^n [A]^m.$$

2.2.4 Inverse Matrix

For squares of dimension n, one can determine the inverse matrix. Thus, the inverse must fulfill the condition:

$$[A]\cdot[A]^{-1}=[A]^{-1}\cdot[A]=[E], \tag{2.12}$$

(if it exists). Calculating this matrix returns to solving a system of linear equations. If $\Delta=\det[A]\neq 0$ (in this case, matrix A is called inversible) we obtain:

$$[A]^{-1}=\frac{1}{\Delta}[A^*], \tag{2.13}$$

where $[A^*]$ is the adjugate of the matrix $[A]$ and Δ is its determinant. The entries of the adjugate are $a^*_{ij}=(-1)^{i+j}\Gamma^{ij}$ and Γ_{ij} are the $n-1$ rank determinants obtained by eliminating line i and column j in the transposed matrix.

For instance, in the case of the matrix

$$[A]=\begin{bmatrix}1 & 3 & 0\\ -1 & 2 & 1\\ 2 & 3 & -1\end{bmatrix}$$

we have:

$$\Delta=-8,\quad [A]^T=\begin{bmatrix}1 & -1 & 2\\ 3 & 2 & 3\\ 0 & 1 & -1\end{bmatrix},\quad \Gamma^{11}=-5;\Gamma^{12}=-3;\Gamma^{13}=3,$$

$$\Gamma^{21}=-1;\Gamma^{22}=-1;\Gamma^{23}=1;\Gamma^{31}=-7;\Gamma^{32}=-3;\Gamma^{33}=5;$$

$$[A^*]=\begin{bmatrix}-5 & 3 & 3\\ 1 & -1 & -1\\ -7 & 3 & 5\end{bmatrix}.$$

It results:

$$[A]^{-1}=\frac{1}{8}\begin{bmatrix}5 & -3 & -3\\ -1 & 1 & 1\\ 7 & -3 & -5\end{bmatrix}=\begin{bmatrix}0.625 & -0.375 & -0.375\\ -0.125 & 0.125 & 0.125\\ 0.875 & -0.375 & -0.625\end{bmatrix}.$$

The rank of an invertible square matrix of size n is n.

2.2.5 Linear Systems

Let us consider the linear system:

$$\begin{aligned} x + 3y + 0z &= 2 \\ -x + 2y + z &= -1 \\ 2x + 3y - z &= 1, \end{aligned} \tag{2.14}$$

which can also be written as:

$$\begin{bmatrix} 1 & 3 & 0 \\ -1 & 2 & 1 \\ 2 & 3 & -1 \end{bmatrix} \begin{Bmatrix} x \\ y \\ z \end{Bmatrix} = \begin{Bmatrix} 2 \\ -1 \\ 1 \end{Bmatrix}. \tag{2.15}$$

The solution can be obtained by inverting the matrix of coefficients, assuming that the determinant of the system is not null:

$$\begin{aligned} \begin{Bmatrix} x \\ y \\ z \end{Bmatrix} &= \begin{bmatrix} 1 & 3 & 0 \\ -1 & 2 & 1 \\ 2 & 3 & -1 \end{bmatrix}^{-1} \begin{Bmatrix} 2 \\ -1 \\ 1 \end{Bmatrix} \\ &= \begin{bmatrix} 0.625 & -0.375 & -0.375 \\ -0.125 & 0.125 & 0.125 \\ 0.875 & -0.375 & -0.625 \end{bmatrix} \begin{Bmatrix} 2 \\ -1 \\ 1 \end{Bmatrix} = \begin{Bmatrix} 1.25 \\ -0.25 \\ 2.75 \end{Bmatrix}. \end{aligned}$$

In general, the system:

$$[A]\{X\} = \{B\}, \tag{2.16}$$

where $[A]$ is a square matrix with a non-zero determinant, has the solution:

$$\{X\} = [A]^{-1}\{B\}. \tag{2.17}$$

2.2.6 Transposed of a Matrix

If $[A]$ is a matrix, its transposed, denoted with $[A]^T$, is the matrix defined by the relations:

$$a'_{ij} = a_{ji}. \tag{2.18}$$

For example, if:

$$[A] = \begin{bmatrix} 1 & -1 & 2 \\ 3 & 2 & 2 \end{bmatrix},$$

then we have:

$$[A]^T = \begin{bmatrix} 1 & 3 \\ -1 & 2 \\ 2 & 2 \end{bmatrix}.$$

In the transposed matrix the lines become columns, and the columns become lines.

2.2.7 Trace of a Matrix

The trace of a matrix A, denoted tr(A) of a square matrix, is the sum of the diagonal elements. If:

$$[A] = \begin{bmatrix} a_{11} & a_{12} & \dots & a_{1n} \\ a_{21} & a_{22} & \dots & a_{2n} \\ \vdots & & & \\ a_{m1} & a_{m2} & \dots & a_{mn} \end{bmatrix},$$

then tr $(A) = a_{11} + a_{22} + a_{33} + \cdots + a_{nn}$. It can be easily demonstrated that:

$$tr([A][B]) = tr([B][A]);$$
$$tr([A]) = tr([A]^T).$$

2.2.8 Matrix Representation of the Cross Product

We attached to two vectors $\bar{b}$ and $\bar{c}$ the column matrices:

$$\{b\} = \begin{Bmatrix} b_x \\ b_y \\ b_z \end{Bmatrix}; \quad \{c\} = \begin{Bmatrix} c_x \\ c_y \\ c_z \end{Bmatrix},$$

and to the vector $\bar{a}$ the skew-symmetric matrix:

$$[a] = \begin{bmatrix} 0 & -a_z & a_y \\ a_z & 0 & -a_x \\ -a_y & a_x & 0 \end{bmatrix}. \tag{2.19}$$

The cross product may be represented as the matrix product:

$$\{c\} = \begin{Bmatrix} c_x \\ c_y \\ c_z \end{Bmatrix} = [a]\{b\} = \begin{bmatrix} 0 & -a_z & a_y \\ a_z & 0 & -a_x \\ -a_y & a_x & 0 \end{bmatrix} \begin{Bmatrix} b_x \\ b_y \\ b_z \end{Bmatrix} = \begin{Bmatrix} a_y b_z - a_z b_y \\ a_z b_x - a_x b_z \\ a_x b_y - a_y b_x \end{Bmatrix}. \tag{2.20}$$

2.3 Eigenvalues and Eigenvectors

In this section the notions of eigenvalues and their eigenvectors are introduced outside the context in which they appear. For each application, the problem of eigenvalues and eigenvectors will be presented separately, highlighting the characteristic properties useful to the application (see Belmann 1969; Dorst 2011; Horn and Johnson 1985; Serre 2010; Staicu 1986).

Let $[A]$ be a square matrix of size n with real elements. It is said that $\{v\}$ is an eigenvector of matrix $[A]$ if we have the relation:

$$[A]\{v\} = \lambda\{v\}. \tag{2.21}$$

In other words, the matrix $[A]$ transforms the vector $\{v\}$ into a collinear one with it. The values λ for which this happens are called the eigenvalues. Relationship (2.21) can also be written as:

$$[A]\{v\} - \lambda[E]\{v\} = \{0\},$$

or

$$([A] - \lambda[E])\{v\} = \{0\}. \tag{2.22}$$

The matrix:

$$[A] - \lambda[E] \tag{2.23}$$

is called the characteristic matrix of the matrix $[A]$. We have:

$$[A] - \lambda[E] = \begin{bmatrix} a_{11} - \lambda & a_{12} & \dots & a_{1n} \\ a_{21} & a_{22} - \lambda & & a_{2n} \\ \vdots & & & \\ a_{n1} & a_{n2} & & a_{nn} - \lambda \end{bmatrix}. \tag{2.24}$$

The characteristic matrix determinant is a polynomial of degree n in λ. The polynomial $\det([A]-\lambda[E])$ is called the characteristic polynomial of the matrix $[A]$. The roots of this polynomial are called the characteristic roots of the matrix.

Finding the vector in (2.21) returns to solve the homogeneous linear system (2.22). It has non-zero solutions if and only if:

$$\det([A] - \lambda[E]) = 0, \tag{2.25}$$

that is, the characteristic polynomial is zero. By solving the Eq. (2.25) we obtain the λ solutions. Returning to the system (2.22) results the eigenvector vectors. If the n values are distinct, there will generally exist n linear independent eigenvectors. In the following it will be assumed that the values are distinct. The case of multiple eigenvalues introduces some computational problems, because the system becomes multiple indefinite, but the results remain basically valid. Some of such cases will be handled within applications.

Consider now $[B]$ similar with $[A]$, namely:

$$[B] = [Q]^{-1}[A][Q], \tag{2.26}$$

with $[Q]$ being non-singular. The similar matrices also have the same characteristic polynomial and the same characteristic roots. Because:

$$\det[Q]^{-1} = \frac{1}{\det[Q]}, \tag{2.27}$$

we have:

$$\begin{aligned} \det([B] - \lambda[E]) &= \det\big([Q]^{-1}[A][Q] - \lambda[E]\big) = \det\big([Q]^{-1}([A] - \lambda[E])[Q]\big) \\ &= \det[Q]^{-1} \det([A] - \lambda[E]) \det[Q] = \det([A] - \lambda[E]). \end{aligned}$$

The two matrices have the same eigenvectors and eigenvalues.

If $[B]$ has the form:

$$[B] = \begin{bmatrix} b_1 & & & 0 \\ & b_2 & & \\ & & \ddots & \\ 0 & & & b_n \end{bmatrix},$$

so with different elements of zero only diagonally we say that $[B]$ is the canonical form.

2.4 Ortogonal Matrix

A square matrix that has the property:

$$[R][R]^T = [R]^T[R] = [E], \tag{2.28}$$

is called an orthogonal matrix. Since the inverse of a matrix fulfills the condition:

$$[R][R]^{-1} = [R]^{-1}[R] = [E], \tag{2.29}$$

it results that the orthogonal matrices have the property:

$$[R]^{-1} = [R]^T. \tag{2.30}$$

This is very important when dealing with numerical problems, since the transposition of a matrix is an extremely simple operation, while the inversion of a matrix requires laborious calculations.

Taking into account the property of the determinants

$$\det([A][B]) = \det[A]\det[B] \text{ and } \det[R]^T = \det[R],$$

it results:

$$\begin{aligned} \det([R][R]^T) &= \det[E] = 1 \\ \det[R]\det[R]^T &= (\det[R])^2 = 1 \\ \det[R] &= \pm 1. \end{aligned} \tag{2.31}$$

2.5 Some Properties of Matrix Operations

We enumerate some of the basic properties of matrix operations that will be widely used in the further development of the book:

- Any matrix can be written as the sum of a symmetric matrix and a skew-symmetric matrix.

Proof Suppose that we can write the matrix $[A]$ as the sum of a symmetric matrix and a skew-symmetric matrix:

$$[A] = [A_s] + [A_a]. \tag{2.32}$$

Let us transpose the written relationship:

$$[A]^T = [A_s]^T + [A_a]^T = [A_s] - [A_a].$$

If we look at the two equations as a matrix system with the unknowns $[A_s]$ and $[A_a]$, we obtain:

$$[A_s] = \frac{1}{2}\left([A] + [A]^T\right) \tag{2.33a}$$

$$[A_a] = \frac{1}{2}\left([A] - [A]^T\right). \tag{2.33b}$$

Written relationships also allow the release of the following two properties:

- The sum between a square matrix and its transposed is a symmetric matrix;
- The difference between a matrix and its transposed is a skew-symmetric matrix.

Example

$$[A] = \begin{bmatrix} 1 & 2 & 3 & 4 \\ 5 & 6 & 7 & 8 \\ 9 & 10 & 11 & 12 \\ 13 & 14 & 15 & 16 \end{bmatrix}, \quad [A]^T = \begin{bmatrix} 1 & 5 & 9 & 13 \\ 2 & 6 & 10 & 14 \\ 3 & 7 & 11 & 15 \\ 4 & 8 & 12 & 16 \end{bmatrix}$$

$$[A_s] = \frac{1}{2}\left([A] + [A]^T\right) = \begin{bmatrix} 1 & 3.5 & 6 & 8.5 \\ 3.5 & 6 & 8.5 & 11 \\ 6 & 8.5 & 11 & 13.5 \\ 8.5 & 11 & 13.5 & 16 \end{bmatrix}$$

$$[A_a] = \frac{1}{2}\left([A] - [A]^T\right) = \begin{bmatrix} 0 & -1.5 & -3 & -4.5 \\ 1.5 & 0 & -1.5 & -3 \\ 3 & 8.5 & 0 & -1.5 \\ 4.5 & 3 & 1.5 & 0 \end{bmatrix}.$$

- The product of a square matrix with its transposed is a symmetric matrix;
- If $[A]$ is symmetric, we have:

$$[A]^T = [A]. \tag{2.34}$$

- If $[A]$ is skew-symmetric, we have

$$[A]^T = -[A], \tag{2.35}$$

or

$$[A] + [A]^T = 0,$$

a relationship that can define a skew-symmetric matrix.

- If $[A]$ is a symmetric matrix, then $[A]^{-1}$ is also symmetric.

$$([A] + [B])^T = [A]^T + [B]^T \tag{2.36}$$
$$([A][B])^T = [B]^T [A]^T \tag{2.37}$$
$$([A][B])^{-1} = [B]^{-1}[A]^{-1} \tag{2.38}$$
$$\left([A]^T\right)^{-1} = \left([A]^{-1}\right)^T \tag{2.39}$$
$$\left([A]^T\right)^T = [A] \tag{2.40}$$
$$\left([A]^{-1}\right)^{-1} = [A] \tag{2.41}$$
$$(k[A])^{-1} = k^{-1}[A]^{-1} \tag{2.42}$$
$$\det [A]^{-1} = (\det[A])^{-1}. \tag{2.43}$$

Applications:

1. If $[A]$ is a symmetric matrix, then $[B] = [T]^T [A][T]$ is a symmetric matrix, where $[T]$ is an arbitrary matrix.
2. Demonstrate:

$$\det [A] = \det [A]^T.$$

3. Let us show that the product of the two orthogonal matrices is also an orthogonal matrix.
4. If $[A]$ is an arbitrary matrix then $[B] = [A][A]^T = [A]^T [A]$ is a symmetrical matrix.
5. Note $[AB] = [A][B] - [B][A]$ (Jacobi's parenthesis). Demonstrate:

$$[[A_1][B_1 C]] + [[B_1][C_1 A]] + [[C_1][A_1 B]] = 0.$$

2.6 Block Matrix

Consider matrices whose elements are all matrices. This approach is useful if we want a compact writing for certain types of equations or we want to partition the matrices with which we work in significant elements of the analysis being performed. A partitioned matrix or a block matrix is a part of a matrix having been broken into some small parts called submatrices or blocks.

Let us consider a matrix of the form:

$$[A] = \begin{bmatrix} a_{11} & a_{12} & a_{13} & a_{14} \\ a_{21} & a_{22} & a_{23} & a_{24} \\ a_{31} & a_{32} & a_{33} & a_{34} \\ a_{41} & a_{42} & a_{43} & a_{44} \end{bmatrix}.$$

We present two ways in which we can partition the given matrix in blocks. We have:

$$[A] = \begin{bmatrix} A_{11} & A_{12} \\ A_{21} & A_{22} \end{bmatrix}$$

with partitioning:

$$[A_{11}] = \begin{bmatrix} a_{11} & a_{12} \\ a_{21} & a_{22} \end{bmatrix}; \quad [A_{12}] = \begin{bmatrix} a_{13} & a_{14} \\ a_{23} & a_{24} \end{bmatrix};$$

$$[A_{21}] = \begin{bmatrix} a_{31} & a_{32} \\ a_{41} & a_{42} \end{bmatrix}; \quad [A_{22}] = \begin{bmatrix} a_{33} & a_{34} \\ a_{43} & a_{44} \end{bmatrix},$$

or partitioning:

$$[A_{11}] = \begin{bmatrix} a_{11} & a_{12} & a_{13} \\ a_{21} & a_{22} & a_{23} \\ a_{31} & a_{32} & a_{33} \end{bmatrix}; \quad [A_{12}] = \begin{bmatrix} a_{14} \\ a_{24} \\ a_{34} \end{bmatrix};$$

$$[A_{21}] = \begin{bmatrix} a_{41} & a_{42} & a_{43} \end{bmatrix}; \quad [A_{22}] = [a_{44}].$$

The sum and the product of partitioned matrices are made according to the same rules as previously defined, as though they were scalar elements, provided that the matrix products so formed were defined (the required conditions for the dimensions of the matrix to be multiplied).

2.7 Matrix Function

2.7.1 General Considerations

In the following, by extension of the definitions used in the case of common functions, matrix functions will be defined. Taking into consideration that a matrix function is a more difficult issue, we will gradually introduce different types of matrices, more common in applications. For symmetric matrices because of the diagonal canonical shape, introducing matrices' functions is easier. By extensions of exponentiation, square root extraction or inversion, such functions will be introduced when the variable is a matrix. In the following we will present, without demonstration (to avoid the fragmentation of exposure) some basic results concerning the diagonalization of a matrix.

2.7.2 Diagonalization of Symmetric Matrices

Theorem 1 *Let $[A]$ be a real symmetric matrix. This can be reduced to the diagonal form by an orthogonal transformation, which means that there is an orthogonal matrix $[T]$ such as:*

$$[R]^T[A][R] = \begin{bmatrix} \lambda_1 & & & 0 \\ & \lambda_2 & & \\ & & \ddots & \\ 0 & & & \lambda_n \end{bmatrix},$$

where λ_i are the characteristic roots of the matrix.

Theorem 2 *A necessary and sufficient condition for the existence of an orthogonal T matrix with the property:*

$$[R]^T[A][R] = \begin{bmatrix} \lambda_1 & & & 0 \\ & \lambda_2 & & \\ & & \ddots & \\ 0 & & & \lambda_n \end{bmatrix}, \quad [R]^T[B][R] = \begin{bmatrix} \mu_1 & & & 0 \\ & \mu_2 & & \\ & & \ddots & \\ 0 & & & \mu_n \end{bmatrix},$$

is that matrices A and B commute.

Symetric matrices commute, so they fulfill the hypotheses of Theorem 2.

2.7.2.1 Usual Matrix Functions

Analytical functions. Any real symmetric matrix can be represented as:

$$[A] = [R]\begin{bmatrix} \lambda_1 & & & 0 \\ & \lambda_2 & & \\ & & \ddots & \\ 0 & & & \lambda_n \end{bmatrix}[R]^T, \tag{2.44}$$

where $[R]$ is an orthogonal matrix. From here it follows immediately, by mathematical induction after k:

$$[A]^k = [R]\begin{bmatrix} \lambda_1^k & & & 0 \\ & \lambda_2^k & & \\ & & \ddots & \\ 0 & & & \lambda_n^k \end{bmatrix}[R]^T. \tag{2.45}$$

The result can be demonstrated for any k rational and then, since any analytical function *f(z)* can be written as a series of powers, we can define the analytical function $f([A])$ of the symmetrical matrix as:

$$f([A]) = [R]\begin{bmatrix} f(\lambda_1) & & & 0 \\ & f(\lambda_2) & & \\ & & \ddots & \\ 0 & & & f(\lambda_n) \end{bmatrix}[R]^T. \tag{2.46}$$

Inverse function: Using the above decomposition process it is natural to define the inverse function as:

$$[A]^{-1} = [R]\begin{bmatrix} \lambda_1^{-1} & & & 0 \\ & \lambda_2^{-1} & & \\ & & \ddots & \\ 0 & & & \lambda_n^{-1} \end{bmatrix}[R]^T. \tag{2.47}$$

It is easy to verify that the matrix $[A]^{-1}$ thus defined checks the relationships:

$$[A]^{-1}[A] = [A][A]^{-1} = [E] \tag{2.48}$$

where $[E]$ is the unit matrix.

Application:

- Using this representation, show that $[A]^{-1}$ is unique.

Square root: A positive defined matrix is a generalization of a positive number. Taking into account the form obtained for a matrix to a rational power, it can be written that:

$$[A]^{\frac{1}{2}} = [R]\begin{bmatrix} (\lambda_1)^{\frac{1}{2}} & & & 0 \\ & (\lambda_2)^{\frac{1}{2}} & & \\ & & \ddots & \\ 0 & & & (\lambda_n)^{\frac{1}{2}} \end{bmatrix}[R]^T. \tag{2.49}$$

This matrix is positively defined. Indeed, if the values are positive, then $(\lambda_i)^{\frac{1}{2}}$ will also be positive and $[A]^{\frac{1}{2}}$ will therefore be a positive defined matrix. It can be shown that the square root of a matrix is unique.

References

Bellman, R: Introduction in Matrix Analysis. Tehnica, Bucharest (1969)
Dorst, L., Lasenby, J. (eds.): Guide to geometric algebra in practice. Springer, New York (2011)
Horn, R.A., Johnson, C.R.: Matrix Analysis. Cambridge University Press, Cambridge (1985)
Serre, D.: Matrices: theory and Applications. Springer, New York (2010)
Staicu, S.: Application of the Matrix Calculus in the Solid Mechanics. Academiei, Bucharest (1986)

Chapter 3
Quadratic Forms

3.1 Introduction

In the following paragraphs we will introduce the notion of quadratic forms of two or more variables. These forms can be introduced starting from several types of problems in which they have a central role. However, we consider it more appropriate to define them from the problem of determining the extreme values of two variable functions as presented in Belmann (1969) and Shimura (2010). This is because most of the applications, in which the quadratic form appears, lead to problems of determining extreme values such as moments of inertia, stress, deformations etc.

3.2 Extreme Values of a Real Function of Two Variables

Consider the real function of two variables $f(x, y)$ defined on the rectangle $a_1 \leq x \leq b_1, a_2 \leq y \leq b_2$.

Developing this function at a point $c = (c_1, c_2)$ inside the definition domain into a Taylor's series gives the following expression:

$$
\begin{aligned}
f(x, y) = f(c_1, c_2) &+ (x - c_1)\frac{\partial f}{\partial x}_{|x=c_1, y=c_2} + (y - c_2)\frac{\partial f}{\partial y}_{|x=c_1, y=c_2} \\
&+ \frac{1}{2}(x - c_1)^2 \frac{\partial^2 f}{\partial x^2}_{|x=c_1, y=c_2} + (x - c_1)(y - c_2)\frac{\partial^2 f}{\partial x \partial y}_{|x=c_1, y=c_2} \\
&+ \frac{1}{2}(y - c_2)^2 \frac{\partial^2 f}{\partial y^2}_{|x=c_1, y=c_2} + \cdots
\end{aligned}
\tag{3.1}
$$

S. Vlase et al., *Eigenvalue and Eigenvector Problems in Applied Mechanics*, Advanced Structured Materials 96, https://doi.org/10.1007/978-3-030-00991-5_3

Let us now consider that $c = (c_1, c_2)$ is a stationary point of the function $f(x, y)$ i.e. we have the equations:

$$\left.\frac{\partial f}{\partial x}\right|_{x=c_1, y=c_2} = 0 \quad \text{and} \quad \left.\frac{\partial f}{\partial y}\right|_{x=c_1, y=c_2} = 0. \tag{3.2}$$

The nature of the function $f(x, y)$ in the neighborhood of the point $c = (c_1, c_2)$ is given by the behavior of the expression that contains the quadratic terms of Taylor's development, namely:

$$Q(x, y) = a(x - c_1)^2 + 2b(x - c_1)(y - c_2) + c(y - c_2)^2, \tag{3.3}$$

where we used the following abbreviations:

$$a = \frac{1}{2}\frac{\partial^2 f}{\partial x^2}_{|x=c_1, y=c_2}; \quad 2b = \frac{\partial^2 f}{\partial x \partial y}_{|x=c_1, y=c_2}; \quad c = \frac{1}{2}\frac{\partial^2 f}{\partial y^2}_{|x=c_1, y=c_2}. \tag{3.4}$$

If we use,

$$u = x - c_1; \quad v = y - c_2, \tag{3.5}$$

then $Q(x, y)$ becomes:

$$Q(u, v) = au^2 + 2buv + cv^2. \tag{3.6}$$

An expression of this type (a homogeneous polynomial of degree two) is called the quadratic form in two variables, u and v.

To determine the nature of the extreme values in $u = 0$, $v = 0$ (maximum, minimum, stationary), we will have to study the sign of the function $Q(u, v)$ in the vicinity of this point. The function $Q(u, v)$ is a homogeneous function in the variables u and v. This allows us to simplify the calculation of the determination of the function's $Q(u, v)$ sign in determining the values that this function takes on the circle $u^2 + v^2 = 1$. Indeed, because $Q(ku, kv) = k^2 Q(u, v)$ for any real k, then the system of values taken on the circle $u^2 + v^2 = k^2$ is in a very simple relationship with the values taken on the circle $u^2 + v^2 = 1$.

If $Q(u, v) > 0$ for any values u and v in the vicinity of the point $(0, 0)$, then $f(x, y)$ has a relative minimum in the point $c = (0, 0)$. If $Q(u, v) < 0$ for any values u and v in the vicinity of the point $Q(u, v) < 0$, then it has a relative maximum in the point $(0, 0)$. If $Q(u, v)$ it can get both positive and negative values we can have a stationary saddle type point.

If $Q(u, v) = 0$ for any values u and v in the vicinity of zero, it is necessary to study the sign of the higher order derivatives.

3.3 Conics and Quadrics

Let us considered the following expression:

$$a'x^2 + 2b'xy + c'y^2 + d'x + e'y + 1 = 0. \tag{3.7}$$

By a convenient change of coordinates (translation) and using obvious notations, it can be brought to the form:

$$au^2 + 2buv + cv^2 - 1 = 0. \tag{3.8}$$

This expression represents the equation of a conic, its nature being given by the value of the coefficients a, b, c. Thus, if a and c are not simultaneously zero, we can have an ellipse, a parabola, or a hyperbola. The equation of a conic (3.8) can be rearranged in matrix from:

$$\begin{bmatrix} u & v \end{bmatrix} \begin{bmatrix} a & b \\ b & c \end{bmatrix} \begin{Bmatrix} u \\ v \end{Bmatrix} = 1. \tag{3.9}$$

In a similar way the expression:

$$a'x^2 + b'y^2 + c'z^2 + 2d'xy + 2e'yz + 2f'zx + g'x + h'y + i'z + 1 = 0 \tag{3.10}$$

can be brought, by a convenient change of coordinates (translation) and using obvious notations, to the form:

$$au^2 + bv^2 + ct^2 + 2duv + 2evt + 2ftu + 1 = 0, \tag{3.11}$$

which represents the well-known equation of a quadric. This can be written in matrix form as:

$$\begin{bmatrix} u & v & t \end{bmatrix} \begin{bmatrix} a & d & f \\ d & b & e \\ f & e & c \end{bmatrix} \begin{Bmatrix} u \\ v \\ t \end{Bmatrix} = 1. \tag{3.12}$$

With the help of these quadratic forms, forms, one can describe conics in the two-dimensional space or quadrics in the three-dimensional space.

3.4 Quadratic Forms in a *n*-Dimensional Space

By generalizing the expression of a quadratic form in the n-dimensional space there is obtained:

$$f = \sum_{i,j=1}^{n} a_{ij} x_i x_j, \tag{3.13}$$

which is called the real quadratic form if the coefficients a_{ij} are real and imaginary if the coefficients a_{ij} are imaginary.

Because $x_i, x_j = x_j x_i$ the coefficient of this product $a_{ij} + a_{ji}$ can be written $2a_{ij}$ by taking $a_{ij} + a_{ji}$. Thus, we can write in lieu of $\left(a_{ij} + a_{ji}\right) x_i x_j$ the product $2a_{ij} x_i x_j$ (in these relations the common summation convention is not applicable). The quadratic form can be written in matrix form as:

$$f = \{X\}^T [A] \{X\}, \tag{3.13$'$}$$

where

$$\{X\}^T = \begin{bmatrix} x_1 & x_2 & \cdots & x_n \end{bmatrix},$$

and

$$[A] = \begin{bmatrix} a_{11} & a_{12} & \ldots & a_{1n} \\ a_{21} & a_{22} & \ldots & a_{2n} \\ \vdots & & & \\ a_{n1} & a_{n2} & \ldots & a_{nn} \end{bmatrix}.$$

It should be noted that A is symmetric $\left(a_{ij} = a_{ji}\right)$. For example, if $n = 2$ we have:

$$f = \begin{bmatrix} x_1 & x_2 \end{bmatrix}^T \begin{bmatrix} a_{11} & a_{12} \\ a_{21} & a_{22} \end{bmatrix} \begin{Bmatrix} x_1 \\ x_2 \end{Bmatrix}, \tag{3.14}$$

where $a_{12} = a_{21}$.

The matrix $[A]$ is named the matrix of the quadratic form f and its rank is the rank of the quadratic form.

If rank $[A] = n$, then it is said that the quadratic form is non-singular.

Let us do the linear transformation now:

$$\begin{aligned} x_1 &= t_{11} y_1 + t_{12} y_2 + \cdots + t_{1n} y_n \\ x_2 &= t_{21} y_1 + t_{22} y_2 + \cdots + t_{2n} y_n \\ &\vdots \\ x_n &= t_{n1} y_1 + t_{n2} y_2 + \cdots + t_{nn} y_n, \end{aligned} \tag{3.15}$$

or in a concise form:

$$\{X\} = [T]\{Y\}. \tag{3.15$'$}$$

With this transformation the quadratic form can be written as,

$$f = \{Y\}^T [T]^T [A][T]\{Y\}. \tag{3.16}$$

The formula for transposing the matrix product has been taken into account:

$$\{X\}^T = ([T]\{Y\})^T - \{Y\}^T [T]^T. \tag{3.17}$$

If we note:

$$[B] = [T]^T [A][T], \tag{3.18}$$

then we get the expression of the quadratic form in y_i:

$$f = \{Y\}^T [B]\{Y\}. \tag{3.16$'$}$$

If $[T]$ is a non-singular matrix ($\det[T] \neq 0$) then, according to a known result $[A]$ and $[B]$ have the same rank. Because $[A]$ is symmetric, $[B]$ will be too symmetric.

3.5 Eigenvalues and Eigenvectors for Quadratic Forms

In the following we return to the question presented at the beginning of the chapter, i.e. to determine the conditions for which a quadratic form is positive, negative or undefined (see Demidovici and Maron 1973; Henderson and Luca 2016; Marin and Arabnia 2010; Marin and Öchsner 2018a, b; Wilkinson 1965).

The investigations will be focused on quadratic forms of two variables.

3.5.1 The Conditions for a Quadratic Form to Be Positive

Consider a quadratic form in the two-dimensional space. We write $Q(u, v)$ in (3.6) in the form:

$$Q(u, v) = a\left(u + \frac{bv}{a}\right)^2 + \left(c - \frac{b^2}{a}\right)v^2 \tag{3.19}$$

with the condition $a \neq 0$. It is immediately obvious that the expression $Q(u, v)$ is positive for any values of u and v if the coefficients a, b and c meet the conditions:

$$a > 0 \quad \text{and} \quad c - \frac{b^2}{a} > 0,$$

or,

$$a > 0 \quad \text{and} \quad \begin{vmatrix} a & b \\ b & c \end{vmatrix} > 0. \tag{3.20}$$

These conditions represent a set of necessary and sufficient conditions for the quadratic form to be positive. It is called a defined positive quadratic form.

If $a = 0$ but $c \neq 0$ we can write $Q(u, v)$ in the form:

$$\begin{aligned} Q(u, v) &= u^2\left(a - \frac{b^2}{c}\right) + c\left(v + \frac{bu}{c}\right)^2 \\ &= -u^2\left(\frac{b^2}{c}\right) + c\left(v + \frac{bu}{c}\right)^2 \end{aligned} \tag{3.21}$$

and the set of necessary and sufficient conditions become:

$$c > 0 \quad \text{and} \quad \begin{vmatrix} a & b \\ b & c \end{vmatrix} = -b^2 > 0 \tag{3.22}$$

(impossible).

If $a = c = 0$ the quadratic form is reduced to $2buv$. If $b \neq 0$ it can take both negative and positive values and if $b = 0$ then we do no longer have a quadratic form.

3.5.2 *Lagrange Multipliers*

We present, without proofs, the Lagrange's multiplier method for determining the extremes of a quadratic function. We therefore want to determine the extreme points of the quadratic function presented in (3.6). The method consists essentially in determining the stationary points of a quadratic function:

$$L(u, v) = au^2 + 2buv + cv^2 - \lambda\left(u^2 + v^2 - 1\right) \tag{3.23}$$

on the unit circle $u^2 + v^2 - 1 = 0$. The newly introduced variable λ is named the Lagrange multiplier. To determine the stationary points returns to solve the equations:

$$\frac{\partial L}{\partial u} = 0; \quad \frac{\partial L}{\partial v} = 0; \quad \frac{\partial L}{\partial \lambda} = 0. \tag{3.24}$$

The third equation is the condition for the solution to be on the unit circle. If we do the calculus, the first two conditions lead to the system:

$$\begin{cases} au + bv - \lambda u = 0 \\ bu + cv - \lambda v = 0 \end{cases},$$

or,

$$\begin{bmatrix} a-\lambda & b \\ b & c-\lambda \end{bmatrix} \begin{Bmatrix} u \\ v \end{Bmatrix} = \begin{Bmatrix} 0 \\ 0 \end{Bmatrix}. \tag{3.25}$$

That is a problem of eigenvalues for the matrix,

$$[A] = \begin{bmatrix} a & b \\ b & c \end{bmatrix},$$

respecting the conditions,

$$u^2 + v^2 - 1 = 0.$$

The homogeneous linear system (3.25) has solutions different from zero if and only if the system determinant is zero:

$$\begin{vmatrix} a-\lambda & b \\ b & c-\lambda \end{vmatrix} = 0, \tag{3.26}$$

which leads to the equation:

$$\lambda^2 - (a+c)\lambda + ac - b^2 = 0. \tag{3.26$'$}$$

If the calculus is made, it is found that the system discriminant is:

$$\Delta = (a-c)^2 + 4b^2, \tag{3.27}$$

therefore greater than or equal to zero (this is due to the symmetry of the matrix A), from which it follows that the roots of the equation are real. Unless $a = c$ and $b = 0$, these roots are distinct.

3.5.3 Fundamental Theorems of Quadratic Forms

Next, without demonstration, the fundamental theorems that define the governing characteristic properties of the quadratic forms will be given.

Fundamental theorem of quadratic forms. Any quadratic form can be reduced to a canonical form by a linear, non-singular transformation.

But, for a quadratic form, the canonical form is not unique. For example, let us take for f the following expression,

$$f = 2x_1x_2 - 6x_2x_3 + 2x_3x_1.$$

Then the linear transformation,

$$\begin{aligned} x_1 &= \frac{1}{2}t_1 + \frac{1}{2}t_2 + 3t_3 \\ x_2 &= \frac{1}{2}t_1 - \frac{1}{2}t_2 - 3t_3 \\ x_3 &= t_3, \end{aligned}$$

allows for the canonical form,

$$f = \frac{1}{2}t_1^2 - \frac{1}{2}t_2^2 + 6t_3^2.$$

With the linear transformation,

$$\begin{aligned} x_1 &= t_1 + 3t_2 + 2t_3 \\ x_2 &= t_1 - t_2 - 2t_3 \\ x_3 &= t_2 \end{aligned}$$

the canonical form is obtained as:

$$f = 2t_1^2 + 6t_2^2 - 8t_3^2.$$

Thus, we obtained from two linear transformations two different canonical forms. It is now the question of finding a property that characterizes the given quadratic form and which does not depend on the chosen linear transformation. This is given by the following law:

Law of inertia. The number of positive and negative coefficients is the same, regardless of the linear transformation by which the canonical form is obtained.

A quadratic form is positively defined if all the coefficients of the canonical form are positive.

A positively defined quadratic form can be characterized without being brought into the canonical form. To show this we present two theorems:

Theorem I *A quadratic form f with real coefficients is positively defined if and only if, for all real values of the variables, from which at least one is different from zero, f is positive.*

The second theorem allows the characterization of a quadratic form using the matrix of coefficients. If $[A]$ there is a matrix of a quadratic form f, minors of the order $1,2,\ldots,n$ are a determinant of order $1,2,\ldots,n$ chosen beginning from the upper left corner, i.e. minors:

$$a_{11}, \begin{vmatrix} a_{11} & a_{12} \\ a_{21} & a_{22} \end{vmatrix}, \ldots, \begin{vmatrix} a_{11} & \ldots & a_{1k} \\ \vdots & & \\ a_{k1} & & a_{kk} \end{vmatrix}, \ldots, \begin{vmatrix} a_{11} & a_{12} & a_{1n} \\ a_{21} & & a_{2n} \\ & & \\ a_{n1} & \ldots & a_{nn} \end{vmatrix} \tag{3.28}$$

are called the main minors of the form f.

Theorem 11 *A quadratic form f in n variables with real coefficients is positively defined if and only if all the main minors of the quadratic form are strictly positive.*

For example, the quadratic form:

$$f = 5x^2 + y^2 + 5z^2 + 4xy - 8xz - 4yz,$$

is positively defined because the main minors:

$$5, \quad \begin{vmatrix} 5 & 2 \\ 2 & 1 \end{vmatrix} = 1, \quad \begin{vmatrix} 5 & 2 & -4 \\ 2 & 1 & -2 \\ -4 & -2 & 5 \end{vmatrix} = 1$$

are positive.

3.5.4 *Schur's Theorem*

If $[A]$ and $[B]$ are positively defined then $[C] = [A][B]$ is positively defined.

Proof It has been shown that a symmetrical matrix with real elements can be written in the canonical form:

$$[A] = [T] \begin{bmatrix} \lambda_1 & & & 0 \\ & \lambda_2 & & \\ & & \ddots & \\ 0 & & & \lambda_n \end{bmatrix} [T]^T.$$

Thus, we can have a parametric representation of the matrix elements as:

$$a_{ij} = \sum_{k=1}^{n} \lambda_k t_{ik} t_{kj},$$

where the sizes of the parameters λ_k are positive. Then it can be written:

$$f = \sum_{i,j=1}^{n} a_{ij}b_{ij}x_i y_j = \sum_{i,j=1}^{n} b_{ij}x_i y_j \left(\sum_{k=1}^{n} \lambda_k t_{ik} t_{kj} \right) = \sum_{k=1}^{n} \lambda_k \left(\sum_{i,j=1}^{n} b_{ij} x_i t_{ik} x_j t_{kj} \right).$$

Being a quadratic form in $x_i t_{ik}$ the expression $\sum_{i,j=1}^{n} b_{ij} x_i t_{ik} x_j t_{kj}$ is positive, unless all the terms $x_i t_{ik}$ are zero. But if that is the case, all x_i must be zero.

3.6 Orthogonal Transformations

Between the linear transformations that bring a quadratic form to the diagonal (canonical) form, of particular interests are the orthogonal transformations. For a symmetric matrix the eigenvectors are orthogonal, which will be shown in the following. Exposure is done more broadly to familiarize the reader with orthogonal transformations, which play a central role in applied mechanics.

Consider the vector: $\{X\} = \left[x_1 \; x_2 \; \cdots \; x_n \right]^T$. We propose to look for a linear transformation to bring the quadratic form to the diagonal form and, in addition, to keep the Euclidian norm of a vector. Let us note this transformation with $[R]$. We have:

$$\{Y\} = [R]\{X\}. \tag{3.29}$$

The condition that the Euclidian norm does not change is expressed by:

$$x_1^2 + x_2^2 + \cdots + x_n^2 = y_1^2 + y_2^2 + \cdots + y_n^2, \tag{3.30}$$

or

$$\{X\}^T\{X\} = \{Y\}^T\{Y\}. \tag{3.31}$$

Taking into account (3.14) it results

$$\{X\}^T\{X\} = \{X\}^T[R]^T[R]\{X\},$$

obtaining the condition it has to fulfill $[R]$:

$$[R]^T[R] = [E]. \tag{3.32}$$

Hence, it also results immediately:

$$[R]^T = [R]^{-1}. \tag{3.33}$$

The matrix $[R]$ that satisfies this condition is called the orthogonal matrix. Noting by $\left\{R^i\right\}$ the column i of the matrix we can write:

$$[R] = \left[R^1 \; R^2 \; \cdots \; R^n \right];$$
$$[R]^T = \left[\left\{R^1\right\}^T \left\{R^2\right\}^T \cdots \{R^n\}^T \right].$$

Using these notations, the condition of orthogonality is expressed as:

$$\left\{R^i\right\}^T \left\{R^j\right\} = \delta_{ij}, \tag{3.34}$$

where δ_{ij} is the Kronecker symbol.

It is clear that the product (3.34) represents the scalar product of vectors $\left\{R^i\right\}$ and $\left\{R^j\right\}$ and expresses the ortonormality of these vectors.

Since the determinant of a matrix is equal to the determinant of the transposed matrix, we have:

$$\det\left([R]^T[R]\right) = \det[E] = 1. \tag{3.35}$$

Then:

$$\det[R]^T \det[R] = (\det[R])^2 = 1, \tag{3.36}$$

therefore,

$$\det[R] = \pm 1. \tag{3.37}$$

It results that an orthogonal transformation is non-singular. An orthogonal transformation is termed proper if $\det[R] = 1$ and improper if $\det[R] = -1$.

Let us now ask the question of finding an orthogonal linear transformation so that the quadratic form f is brought into the canonical form. We must have:

$$[B] = [R]^T[A][R], \tag{3.38}$$

where $[B]$ is diagonal and $[R]$ is orthogonal. Multiplying (3.38) from the left-hand side with [R] gives:

$$[R][B] = [A][R]. \tag{3.39}$$

If we denote by r_{ij} the elements of the matrix $[R]$, we obtain n equalities of the form:

$$
\begin{aligned}
r_{1i}b_i &= a_{11}r_{1i} + a_{12}r_{2i} + \cdots + a_{1n}r_{ni} \\
r_{2i}b_i &= a_{21}r_{1i} + a_{22}r_{2i} + \cdots + a_{2n}r_{ni} \\
&\vdots \\
r_{ni}b_i &= a_{n1}r_{1i} + a_{n2}r_{2i} + \cdots + a_{nn}r_{ni} \quad i = \overline{1,n}
\end{aligned}
\tag{3.40}
$$

or

$$
\begin{aligned}
&(a_{11} - b_i)r_{1i} + a_{12}r_{2i} + \cdots + a_{1n}r_{ni} = 0 \\
&a_{21}r_{1i} + (a_{22} - b_i)r_{2i} + \cdots + a_{2n}r_{ni} = 0 \\
&\vdots \\
&a_{n1}r_{1i} + a_{n2}r_{2i} + \cdots + (a_{nm} - b_i)r_{ni} = 0 \quad i = \overline{1,n},
\end{aligned}
\tag{3.41}
$$

which represent a homogeneous linear systems of n unknowns with n equations. They have non-null solutions if and only if:

$$
\begin{vmatrix}
a_{11} - b_i & a_{12} & \cdots & a_{1n} \\
a_{21} & a_{22} - b_i & & a_{2n} \\
\vdots & & & \\
a_{n1} & a_{n2} & & a_{nn} - b_i
\end{vmatrix}
= 0, \quad i = \overline{1,n}
\tag{3.42}
$$

or:

$$
\det([A] - b_i[E]) = 0, \quad i = \overline{1,n}
\tag{3.43}
$$

and allows to find the values b_i as solutions of the algebraic equation of degree n:

$$
P(\lambda) = \det([A] - \lambda[E]) = 0.
\tag{3.44}
$$

If the column i of the matrix $[R]$ is denoted by $\{R^i\}$, the systems (3.41) are written as:

$$
([A] - b_i[E])\{R^i\} = \{0\}.
\tag{3.45}
$$

Thus, if we know b_i, $\{R^i\}$ is determined as a solution of the homogeneous linear system (3.45). The linear transformation $[R] = \left(R^1 \; R^2 \; \cdots \; R^n \right)$ allows the matrix $[A]$ to be brought into diagonal form. It is easy to see that $\{R^i\}$ are the eigenvectors of the matrix $[A]$ and b_i are the eigenvalues.

The system (3.45) is an undetermined linear system. It results for $\{R^i\}$ an infinite number of solutions, consisting of colinear vectors. It can be said that $\{R^i\}$, by extending the terminology from the three-dimensional space, is actually an eigendirection.

Taking into account the relation (3.34), it follows that we have: $\{R^i\}^T\{R^j\} = \delta_{ij}$ (Kronecker delta). The relationship $\{R^i\}^T\{R^i\} = 1$ introduces an additional condition between the components of $\{R^i\}$, i.e. normalizes the vector to be unity. Through this normalization, making an abstraction of sense, $\{R^i\}$ becomes uniquely determined and represents a unit vector. The normalization condition is written as:

$$e_{1i}^2 + e_{2i}^2 + \cdots + e_{ni}^2 = 1, \quad i = \overline{1,n}. \tag{3.46}$$

3.7 Invariants of Quadratic Forms

To simplify the presentation, the results for the three-dimensional space will be expressed, the results for any n-dimensional space being similar. Let us consider a symmetrical quadratic form f with the following coefficient matrix:

$$[A] = \begin{bmatrix} a_{11} & a_{12} & a_{13} \\ a_{21} & a_{22} & a_{23} \\ a_{31} & a_{32} & a_{33} \end{bmatrix}.$$

The characteristic equation of matrix $[A]$ reads:

$$\begin{vmatrix} a_{11}-\lambda & a_{12} & a_{13} \\ a_{21} & a_{22}-\lambda & a_{23} \\ a_{31} & a_{32} & a_{33}-\lambda \end{vmatrix} = 0, \tag{3.47}$$

can be written after development as:

$$\lambda^3 - \lambda^2(a_{11}+a_{22}+a_{33}) \\ + \lambda\left(\begin{vmatrix} a_{11} & a_{12} \\ a_{21} & a_{22} \end{vmatrix} + \begin{vmatrix} a_{22} & a_{23} \\ a_{23} & a_{33} \end{vmatrix} + \begin{vmatrix} a_{11} & a_{13} \\ a_{31} & a_{33} \end{vmatrix}\right) - \begin{vmatrix} a_{11} & a_{12} & a_{13} \\ a_{21} & a_{22} & a_{23} \\ a_{31} & a_{32} & a_{33} \end{vmatrix} = 0. \tag{3.48}$$

With the notations,

$$I_1 = a_{11} + a_{22} + a_{33}, \tag{3.49}$$

$$I_2 = \begin{vmatrix} a_{11} & a_{12} \\ a_{21} & a_{22} \end{vmatrix} + \begin{vmatrix} a_{22} & a_{23} \\ a_{23} & a_{33} \end{vmatrix} + \begin{vmatrix} a_{11} & a_{13} \\ a_{31} & a_{33} \end{vmatrix}, \tag{3.50}$$

$$I_3 = \begin{vmatrix} a_{11} & a_{12} & a_{13} \\ a_{21} & a_{22} & a_{23} \\ a_{31} & a_{32} & a_{33} \end{vmatrix}, \tag{3.51}$$

the characteristics equation becomes:

$$\lambda^3 - I_1\lambda^2 + I_2\lambda - I_3 = 0. \tag{3.47'}$$

If $[T]$ is a linear transformation, then the matrix $[A]$ and the matrix $[B] = [T]^{-1}[A][T]$, according to a previously result, have the same eigenvalues, so the same characteristic polynomial. This results in the invariance of the coefficients I_1, I_2, I_3.

In particular, if we choose for $[T]$ the orthogonal matrix $[R]$, then the obtained matrix will have in this case the same eigenvalues and eigenvectors as the given matrix (as $[R]^T = [R]^{-1}$ it results $[B] = [R]^T[A][R]$).

Hence I_1, I_2, I_3 are invariants of the quadratic form f and they will not depend on the orthogonal coordinate system in which the quadratic form is represented. In particular, the canonical matrix can be:

$$[B] = \begin{bmatrix} \lambda_1 & 0 & 0 \\ 0 & \lambda_2 & 0 \\ 0 & 0 & \lambda_3 \end{bmatrix}$$

and then we have,

$$\begin{aligned} I_1 &= \lambda_1 + \lambda_2 + \lambda_3 \\ I_2 &= \lambda_1\lambda_2 + \lambda_2\lambda_3 + \lambda_3\lambda_1 \\ I_3 &= \lambda_1\lambda_2\lambda_3. \end{aligned}$$

3.8 Examples

A1. Consider the quadratic form:

$$f(x, y) = \begin{bmatrix} x & y \end{bmatrix}\begin{bmatrix} 2 & 1 \\ 1 & 2 \end{bmatrix}\begin{Bmatrix} x \\ y \end{Bmatrix}.$$

The equation $f(x, y) = 1$ is the equation of a conic. We propose to determine the eigenvalues and eigenvectors of the matrix of the quadratic form and to represent the conic attached to this quadratic form.

The problem of the eigenvalues and eigenvectors leads to the system:

$$\left(\begin{bmatrix} 2 & 1 \\ 1 & 2 \end{bmatrix} - \lambda\begin{bmatrix} 1 & 0 \\ 0 & 1 \end{bmatrix}\right)\begin{Bmatrix} e_1 \\ e_2 \end{Bmatrix} = \begin{Bmatrix} 0 \\ 0 \end{Bmatrix}.$$

By applying the zero-determinant condition:

$$\begin{vmatrix} 2-\lambda & 1 \\ 1 & 2-\lambda \end{vmatrix} = 0,$$

the characteristic equation results:

$$(2-\lambda)^2 - 1 = 0,$$

with the solutions: $\lambda_1 = 1$ and $\lambda_2 = 3$.

The canonical form of the coefficients matrix after an orthogonal transformation is:

$$[B] = \begin{bmatrix} 1 & 0 \\ 0 & 3 \end{bmatrix}.$$

For the solution $\lambda_1 = 1$ we obtain the system:

$$\begin{bmatrix} 2-1 & 1 \\ 1 & 2-1 \end{bmatrix} \begin{Bmatrix} e_{11} \\ e_{21} \end{Bmatrix} = \begin{Bmatrix} 0 \\ 0 \end{Bmatrix},$$

with the solution $e_{21} = -e_{11}$. The normalization condition $e_{11}^2 + e_{21}^2 = 1$ leads to the solution:

$$e_{21} = -e_{11} = \pm\frac{\sqrt{2}}{2},$$

and its first eigenvector can be written:

$$\{v_1\} = \pm \begin{Bmatrix} \frac{\sqrt{2}}{2} \\ -\frac{\sqrt{2}}{2} \end{Bmatrix}.$$

The second solution $\lambda_2 = 3$ yields:

$$\begin{bmatrix} 2-3 & 1 \\ 1 & 2-3 \end{bmatrix} \begin{Bmatrix} e_{12} \\ e_{22} \end{Bmatrix} = \begin{Bmatrix} 0 \\ 0 \end{Bmatrix},$$

with the solution $e_{12} = e_{22}$ which, together with the normalization condition $e_{12}^2 + e_{22}^2 = 1$, will give:

$$e_{12} = e_{22} = \pm\frac{\sqrt{2}}{2},$$

so its second eigenvector can be written,

$$\{v_2\} = \pm \left\{ \begin{array}{c} \frac{\sqrt{2}}{2} \\ \frac{\sqrt{2}}{2} \end{array} \right\}.$$

Applying the condition that the two orthogonal vectors form the basis of a rectangular axis, it results:

$$\{v_1\} = \left\{ \begin{array}{c} \frac{\sqrt{2}}{2} \\ -\frac{\sqrt{2}}{2} \end{array} \right\}; \quad \{v_2\} = \left\{ \begin{array}{c} \frac{\sqrt{2}}{2} \\ \frac{\sqrt{2}}{2} \end{array} \right\} \quad \text{or}$$

$$\{v_1\} = -\left\{ \begin{array}{c} \frac{\sqrt{2}}{2} \\ -\frac{\sqrt{2}}{2} \end{array} \right\}; \quad \{v_2\} = -\left\{ \begin{array}{c} \frac{\sqrt{2}}{2} \\ \frac{\sqrt{2}}{2} \end{array} \right\}.$$

In these cases $\det[R] = 1$. The rotation matrix built with the eigenvectors is:

$$[R] = [\{v_1\}\{v_2\}] = \pm \begin{bmatrix} \frac{\sqrt{2}}{2} & \frac{\sqrt{2}}{2} \\ -\frac{\sqrt{2}}{2} & \frac{\sqrt{2}}{2} \end{bmatrix}.$$

The variables x and y are related to the canonical variables X and Y by the transformation:

$$\left\{ \begin{array}{c} x \\ y \end{array} \right\} = \pm \begin{bmatrix} \frac{\sqrt{2}}{2} & \frac{\sqrt{2}}{2} \\ -\frac{\sqrt{2}}{2} & \frac{\sqrt{2}}{2} \end{bmatrix} \left\{ \begin{array}{c} X \\ Y \end{array} \right\}.$$

It is easy to check that:

$$\begin{bmatrix} 1 & 0 \\ 0 & 3 \end{bmatrix} = \begin{bmatrix} \frac{\sqrt{2}}{2} & -\frac{\sqrt{2}}{2} \\ \frac{\sqrt{2}}{2} & \frac{\sqrt{2}}{2} \end{bmatrix} \begin{bmatrix} 2 & 1 \\ 1 & 2 \end{bmatrix} \begin{bmatrix} \frac{\sqrt{2}}{2} & \frac{\sqrt{2}}{2} \\ -\frac{\sqrt{2}}{2} & \frac{\sqrt{2}}{2} \end{bmatrix},$$

and the canonical form is:

$$f = X^2 + 3Y^2.$$

The relationship obtained is the equation of a conic with the center in the origin, the eigenvalues determine the semi-minor and semi-major axis of the conic and the eigenvectors will represent the unit vectors of the coordinate system in which the conic has the canonic form (Fig. 3.1).

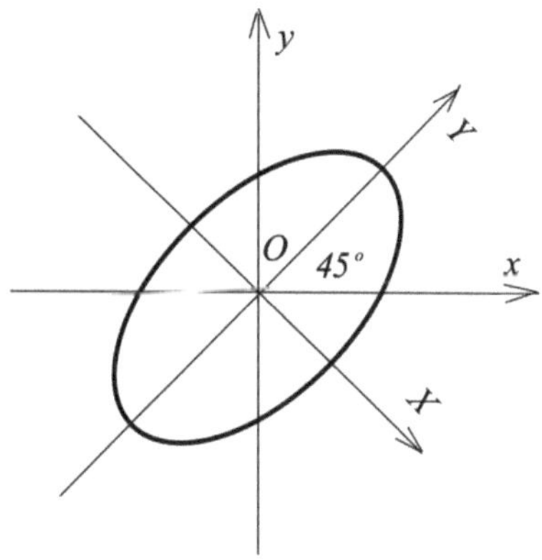

Fig. 3.1 The ellipse of a quadratic form

Since $\frac{\sqrt{2}}{2} = \sin\frac{\pi}{4} = \cos\frac{\pi}{4}$ the matrix $[R]$ can be written:

$$[R] = \pm\begin{bmatrix} \cos\frac{\pi}{4} & \sin\frac{\pi}{4} \\ -\sin\frac{\pi}{4} & \cos\frac{\pi}{4} \end{bmatrix}$$

and its physical signification is a rotation of the coordinate system in a plane with the angle $\frac{3\pi}{4}$ or $-\frac{\pi}{4}$. Performing this rotation, the conic equation is brought into the canonical form.

A2. Consider the quadratic form:

$$f = [x\ y\ z]\begin{bmatrix} 4 & -1 & 0 \\ -1 & 5 & -1 \\ 0 & -1 & 4 \end{bmatrix}\begin{Bmatrix} x \\ y \\ z \end{Bmatrix}.$$

We propose to determine the eigenvectors and the eigenvalues of the quadratic form and to build the quadric associated with this form.

The problem of eigenvalues and eigenvectors leads to the system:

$$\left(\begin{bmatrix} 4 & -1 & 0 \\ -1 & 5 & -1 \\ 0 & -1 & 4 \end{bmatrix} - \lambda\begin{bmatrix} 1 & 0 & 0 \\ 0 & 1 & 0 \\ 0 & 0 & 1 \end{bmatrix}\right)\begin{Bmatrix} e_1 \\ e_2 \\ e_3 \end{Bmatrix} = \begin{Bmatrix} 0 \\ 0 \\ 0 \end{Bmatrix}.$$

Its solution in increasing order is as follows:

$$\begin{vmatrix} 4-\lambda & -1 & 0 \\ -1 & 5-\lambda & -1 \\ 0 & -1 & 4-\lambda \end{vmatrix} = 0,$$

$$(4-\lambda)\left(\lambda^2 - 9\lambda - 18\right) = 0$$

with the solutions, in an increasing order,

$$\lambda_1 = 3; \quad \lambda_2 = 4; \quad \lambda_3 = 6.$$

The canonical form of the coefficients matrix will be:

$$[B] = \begin{bmatrix} 3\,0\,0 \\ 0\,4\,0 \\ 0\,0\,6 \end{bmatrix}.$$

For $\lambda_1 = 3$, the following system is obtained:

$$\begin{bmatrix} 4-3 & -1 & 0 \\ -1 & 5-3 & -1 \\ 0 & -1 & 4-3 \end{bmatrix} \begin{Bmatrix} e_{11} \\ e_{21} \\ e_{31} \end{Bmatrix} = \begin{Bmatrix} 0 \\ 0 \\ 0 \end{Bmatrix},$$

or,

$$\begin{aligned} & e_{11} - e_{21} = 0, \\ & -e_{11} + 2e_{21} - e_{31} = 0, \\ & -e_{21} + e_{31} = 0. \end{aligned}$$

The rank of the system is 2 and only two equations can be used:

$$\begin{aligned} & e_{11} - e_{21} = 0 \\ & -e_{21} - e_{31} = 0. \end{aligned}$$

The solution $e_{11} = e_{21} = e_{31}$ is obtained. The normalization condition,

$$e_{11}^2 + e_{21}^2 + e_{31}^2 = 1,$$

leads to $3e_{11}^2 = 1$, from where:

$$e_{11} = \pm\frac{\sqrt{3}}{3}.$$

We choose the positive solution:

$$e_{11} = e_{21} = e_{31} = \frac{\sqrt{3}}{3},$$

and then the first eigenvector will be:

$$\{v_1\} = \frac{\sqrt{3}}{3}\begin{Bmatrix} 1 \\ 1 \\ 1 \end{Bmatrix}.$$

For $\lambda_2 = 4$ we obtain,

$$\begin{bmatrix} 4-4 & -1 & 0 \\ -1 & 5-4 & -1 \\ 0 & -1 & 4-4 \end{bmatrix}\begin{Bmatrix} e_{12} \\ e_{22} \\ e_{32} \end{Bmatrix} = \begin{Bmatrix} 0 \\ 0 \\ 0 \end{Bmatrix},$$

or,

$$\begin{aligned} &e_{22} = 0, \\ &-e_{12} + e_{22} - e_{32} = 0. \end{aligned}$$

It results $e_{32} = -e_{12}$ and the use of the normalization condition leads to $2e_{12}^2 = 1$, where,

$$e_{12} = \pm\frac{\sqrt{2}}{2} = -e_{32}.$$

We choose:

$$e_{12} = \frac{\sqrt{2}}{2} \quad \text{and} \quad e_{32} = -\frac{\sqrt{2}}{2},$$

and then the second eigenvector is:

$$\{v_2\} = \frac{\sqrt{2}}{2}\begin{Bmatrix} 1 \\ 0 \\ -1 \end{Bmatrix}.$$

For $\lambda_3 = 6$ we get:

$$\begin{bmatrix} 4-6 & -1 & 0 \\ -1 & 5-6 & -1 \\ 0 & -1 & 4-6 \end{bmatrix}\begin{Bmatrix} e_{13} \\ e_{23} \\ e_{33} \end{Bmatrix} = \begin{Bmatrix} 0 \\ 0 \\ 0 \end{Bmatrix},$$

or,

$$\begin{aligned}
&-2e_{13} - e_{23} = 0,\\
&-e_{13} - e_{23} - e_{33} = 0,\\
&-e_{23} - 2e_{33} = 0.
\end{aligned}$$

It results:

$$e_{13} = -\frac{1}{2}e_{23} = e_{33}.$$

The normalization condition offers: $6e_{13}^2 = 1$ from where: $e_{13} = \pm\frac{\sqrt{6}}{6}$. We select the positive sign imposing the condition that the orthogonal coordinate system formed with the vectors $\vec{v}_1$, $\vec{v}_2$, $\vec{v}_3$ is rectangular.

Or alternatively expressed: knowing $\vec{v}_1$, $\vec{v}_2$ the third unit vector is determined by the condition that $\vec{v}_1$, $\vec{v}_2$, $\vec{v}_3$ define an orthogonal coordinate system. In this case $\vec{v}_3$ would result from the cross product:

$$\vec{v}_3 = \vec{v}_1 \times \vec{v}_2.$$

By any of the methods it results:

$$\{v_3\} = \frac{\sqrt{6}}{6}\begin{Bmatrix} 1 \\ -2 \\ 1 \end{Bmatrix}.$$

The matrix of rotation has as columns the three determined vectors:

$$[R] = \Big[\{v_1\}\ \{v_2\}\ \{v_3\}\Big] = \begin{bmatrix} \frac{\sqrt{3}}{3} & \frac{\sqrt{2}}{2} & \frac{\sqrt{6}}{6} \\ \frac{\sqrt{3}}{3} & 0 & -\frac{2\sqrt{6}}{6} \\ \frac{\sqrt{3}}{3} & -\frac{\sqrt{2}}{2} & \frac{\sqrt{6}}{6} \end{bmatrix}.$$

The variables x, y, z are related to the canonical variables X, Y, Z by the following relation:

$$\begin{Bmatrix} x \\ y \\ z \end{Bmatrix} = \begin{bmatrix} \frac{\sqrt{3}}{3} & \frac{\sqrt{2}}{2} & \frac{\sqrt{6}}{6} \\ \frac{\sqrt{3}}{3} & 0 & -\frac{2\sqrt{6}}{6} \\ \frac{\sqrt{3}}{3} & -\frac{\sqrt{2}}{2} & \frac{\sqrt{6}}{6} \end{bmatrix}\begin{Bmatrix} X \\ Y \\ Z \end{Bmatrix}.$$

It is easy to check that:

$$[R]^T = [A][R] = \begin{bmatrix} 3 & 0 & 0 \\ 0 & 4 & 0 \\ 0 & 0 & 6 \end{bmatrix},$$

and then the quadratic form results in canonical coordinates:

$$f = \begin{bmatrix} X & Y & Z \end{bmatrix} \begin{bmatrix} 3 & 0 & 0 \\ 0 & 4 & 0 \\ 0 & 0 & 6 \end{bmatrix} \begin{Bmatrix} X \\ Y \\ Z \end{Bmatrix}$$
$$= 3X^2 + 4Y^2 + 6Z^2 = \frac{X^2}{\left(\frac{1}{\sqrt{3}}\right)^2} + \frac{Y^2}{\left(\frac{1}{2}\right)^2} + \frac{Z^2}{\left(\frac{1}{\sqrt{6}}\right)^2},$$

which represents the equation of an ellipsoid. The corresponding eigenvectors will represent the unit vectors of the coordinate system in which the quadric has the canonical form.

Note that, as for conics, the rotation matrix that ensures the transformation to the same canonical form is not unique.

A3. Consider a quadratic form defined by the matrix:

$$[A] = \begin{bmatrix} 3 & 5 & 2 & 6 & 7 & 5 \\ 5 & 4 & 6 & 2 & 4 & 7 \\ 2 & 6 & 4 & 3 & 2 & 5 \\ 6 & 2 & 3 & 5 & 3 & 6 \\ 7 & 4 & 2 & 3 & 4 & 1 \\ 5 & 7 & 5 & 6 & 1 & 3 \end{bmatrix}.$$

The eigenvalues, obtained using a numerical procedure are:

$$[\lambda] = [-4.9232; -5.5521; -0.9162; 3.2754; 5.6520; 25.4642]$$

and the corresponding eigenvectors are ordered, on columns, in the matrix:

$$[R] = \begin{bmatrix} -0.7039 & 0.3006 & -0.2139 & -0.0197 & -0.4136 & 0.4438 \\ 0.4252 & 0.4552 & -0.3948 & -0.3987 & 0.3113 & 0.4483 \\ -0.2777 & 0.0194 & 0.7084 & -0.1734 & 0.5124 & 0.3577 \\ 0.3830 & 0.2447 & 0.2667 & 0.7165 & -0.2110 & 0.4055 \\ 0.3046 & -0.4204 & 0.2562 & -0.4742 & -0.5703 & 0.3388 \\ -0.0846 & -0.6822 & -0.3997 & 0.2702 & 0.3158 & 0.4415 \end{bmatrix}.$$

The quadratic form becomes in canonical coordinates:

$$f = -4.9232X_1^2 - 5.5521X_2^2 - 0.9162X_2^3 + 3.2754X_4^2 + 5.6520X_5^2 + 25.4642X_6^2.$$

A4. Consider a quadratic form defined by the matrix:

$$[A] = \begin{bmatrix} 2 & 1 & 1 & 1 & 1 \\ 2 & 3 & 2 & 2 & 2 \\ 3 & 3 & 4 & 3 & 3 \\ 4 & 4 & 4 & 5 & 4 \\ 5 & 5 & 5 & 5 & 6 \end{bmatrix}.$$

The eigenvalues, obtained using a numerical procedure are: $\lambda_1 = 16$; $\lambda_2 = \lambda_3 = \lambda_4 = \lambda_5 = 1$, they have a multiple (quadruple) root. The homogeneous linear system that offers the eigenvalues is multiple undetermined and offers more choice options, all equally good. A solution for the eigenvector matrix is:

$$[R] = \begin{bmatrix} -0.8854 & 0.1348 & 0.7195 & -0.7213 & 0.7890 \\ 0.1265 & 0.2697 & 0.2959 & -0.2866 & -0.5083 \\ 0.1897 & 0.4045 & -0.2778 & 0.2864 & 0.1411 \\ 0.2530 & 0.5394 & -0.5197 & 0.5269 & -0.2826 \\ 0.3162 & 0.6742 & -0.2179 & 0.1946 & -0.1392 \end{bmatrix}$$

The canonical form of the matrix in the coordinate system defined by the eigenvectors is:

$$[B] = [R]^T[A][R] = \begin{bmatrix} 16 & 0 & 0 & 0 & 0 \\ 0 & 1 & 0 & 0 & 0 \\ 0 & 0 & 1 & 0 & 0 \\ 0 & 0 & 0 & 1 & 0 \\ 0 & 0 & 0 & 0 & 1 \end{bmatrix}$$

and the quadratic form becomes, expressed in the canonical coordinates:

$$f = 16X_1^2 + X_2^2 + X_3^2 + X_4^2 + X_5^2.$$

To determine the eigenvalues and eigenvectors, the EIG subroutine in the MatLab library was used. If E = EIG(X) is used, we obtain for E a vector that contains the eigenvalues of the square matrix X. If we call the routine in the form [V, D] = EIG(X) we obtain two square matrices, D—a diagonal array with the eigenvalues and V whose columns are the eigenvectors corresponding to their eigenvalues in matrix D. There is the relation: X * V = V * D.

References

Bellman, R.: Introduction in matrix analysis. Ed. Tehnica, Bucharest (1969)
Demidovici, B., Maron, I.: Elements de calcul numerique. Editions Moscou (1973)
Henderson, J., Luca, R.: Boundary Value Problems for Systems of Differential, Difference and Fractional Equations: Positive Solutions. Elsevier, New York (2016)
Marin, M., Arabnia, H.: Equations of Evolution. Elliot & Fitzpatrick, Athens (2010)
Marin, M., Öchsner, A.: Complements of Higher Mathematics. Springer, Cham (2018a)
Marin, M., Öchsner, A.: Essentials of Partial Differential Equations. Springer, Cham (2018b)
Shimura, G.: Arithmetic of Quadratic Forms. Springer, Berlin (2010)
Wilkinson, J.H.: The Algebraic Eigenvalue Problem. Clarendon Press, Oxford (1965)

Chapter 4
Rigid Body Mechanics

4.1 Finite Rotation

4.1.1 Defining the Position of a Rigid Body

Knowing the position of a rigid body means to know the position of all its points. The rigid behavior condition makes the number of parameters required to define the position in space to be six. For a better understanding of the notions introduced in this paragraph, we will first study a particular case, namely, the case of plane rotations. Thus, in the plan, we can determine the position of a point belonging to a rigid body attaching to the body an Oxy local coordinate system, which is solidary with the rigid body. This is defined, with respect to the global coordinate system, by the coordinates of the origin $O(x_0, y_0)$ and by the angle θ made by the Ox axis with the axis Ox_1 (see Fig. 4.1). If we know the position of a point $M(x, y)$ with respect to the Oxy local reference system, its position with respect to the global reference system $O_1x_1y_1$ is determined with the relations:

$$\begin{Bmatrix} x_1 \\ y_1 \end{Bmatrix} = \begin{bmatrix} \cos\theta & -\sin\theta \\ \sin\theta & \cos\theta \end{bmatrix} \begin{Bmatrix} x \\ y \end{Bmatrix}, \tag{4.1}$$

which is well-known in analytical geometry. With notations

$$\{r_1\} = \begin{Bmatrix} x_1 \\ y_1 \end{Bmatrix}; \quad \{r_0\} = \begin{Bmatrix} x_0 \\ y_0 \end{Bmatrix}; \quad \{r\} = \begin{Bmatrix} x \\ y \end{Bmatrix}; \quad [\mathbf{R}] = \begin{bmatrix} \cos\theta & -\sin\theta \\ \sin\theta & \cos\theta \end{bmatrix} \tag{4.2}$$

we have:

$$\{r_1\} = \{r_0\} + \{r\} = \{r_0\} + [\mathbf{R}]\{\rho\}. \tag{4.3}$$

S. Vlase et al., *Eigenvalue and Eigenvector Problems in Applied Mechanics*, Advanced Structured Materials 96, https://doi.org/10.1007/978-3-030-00991-5_4

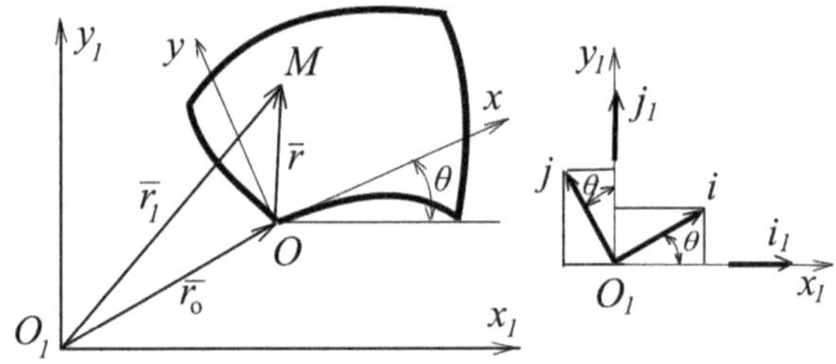

Fig. 4.1 Local and global coordinate systems in a plane

The same relationships could be achieved by writing:

$$\bar{r}_1 = \bar{r}_0 + \bar{r} = x_{01}\bar{i}_1 + y_{01}\bar{j}_1 + x\bar{i} + y\bar{j}. \tag{4.4}$$

Considering that (see Fig. 4.1b):

$$\{i\} = \begin{Bmatrix} \cos\theta \\ \sin\theta \end{Bmatrix}; \quad \{j\} = \begin{Bmatrix} -\sin\theta \\ \cos\theta \end{Bmatrix} \tag{4.5}$$

the following relation is obtained:

$$\{r_1\} = \{r_0\} + [\mathbf{R}]\{\rho\}, \tag{4.6}$$

where

$$[\mathbf{R}] = \begin{bmatrix} \cos\theta & -\sin\theta \\ \sin\theta & \cos\theta \end{bmatrix} = [\{i\}\ \{j\}], \tag{4.7}$$

represents the rotation matrix whose columns contain the unit vectors of the local reference system. It is easy to verify that $[\mathbf{R}]$ is an orthonormal matrix. Thus,

$$[\mathbf{R}][\mathbf{R}]^T = [\mathbf{R}]^T[\mathbf{R}] = [\mathbf{E}] \tag{4.8a}$$

and

$$\det[\mathbf{R}] = 1. \tag{4.8b}$$

The position of the Oxy local coordinate system that allows us to determine the global coordinates of a certain point $M(x, y)$ is defined by three scalar parameters: the coordinates of the origin of this system with respect to the global reference system and the angle of rotation θ of the local reference system towards the fixed reference system.

In an analogous way, consider a rigid body to which we attach an $Oxyz$ three-dimensional local coordinate system. The position of the material point M related

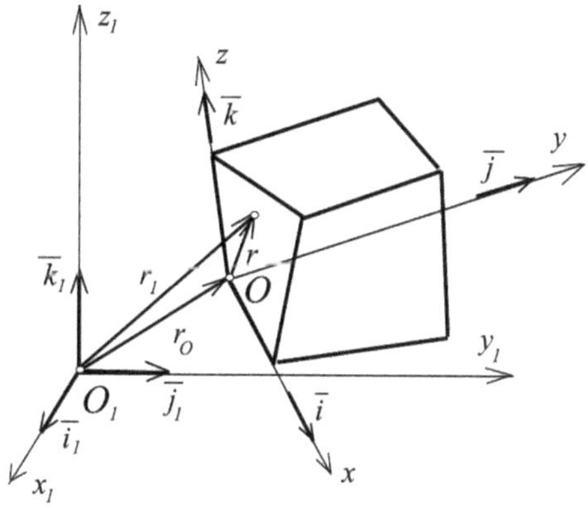

Fig. 4.2 Three-dimensional global and local coordinate systems

to the local coordinate system is known by its coordinates x, y, z. The position of M related to the global coordinate system $O_1x_1y_1z_1$ is determined if we know:

- the position of the origin of the local coordinate system O;
- the unit vectors $\vec{i}$, $\vec{j}$ and $\vec{k}$ corresponding to the three axes.

As a vector has three components, the knowledge of the three unit vectors would require the knowledge of nine scalar components. The conditions that $\vec{i}$, $\vec{j}$ and $\vec{k}$, represent the basis of an orthonormal coordinate system, leads to the following six conditions:

$$\begin{aligned} \bar{i}^2 = 1, \quad \bar{j}^2 = 1, \quad \bar{k}^2 = 1, \\ \bar{i} \cdot \bar{j} = 0, \quad \bar{j} \cdot \bar{k} = 0, \quad \bar{k} \cdot \bar{i} = 0. \end{aligned} \tag{4.9}$$

Thus, three scalar parameters are enough to characterize the position of the local coordinate system axes. If we consider in addition the three parameters defining the position of the origin O, it results that the position of the local coordinate system, and thus of the rigid body, is defined by six scalar parameters (Fig. 4.2).

The unit vectors $\bar{i}$, $\bar{j}$ and $\bar{k}$ are defined by their direction cosines (or directional cosines):

$$\{i\} = \begin{Bmatrix} \cos(x, x_1) \\ \cos(x, y_1) \\ \cos(x, z_1 \end{Bmatrix}, \quad \{j\} = \begin{Bmatrix} \cos(y, x_1) \\ \cos(y, y_1) \\ \cos(y, z_1 \end{Bmatrix}, \quad \{k\} = \begin{Bmatrix} \cos(z, x_1) \\ \cos(z, y_1) \\ \cos(z, z_1 \end{Bmatrix}, \tag{4.10}$$

where $\cos(x, x_1)$ represents the direction cosine (cosine of the angle between the Ox-axis and O_1x_1—the projection of the unit vector $\bar{i}$ on the O_1x_1 axis) etc. We have the relationship

$$\{r_1\} = \{r_O\} + \{r\},$$

or in components

$$x_1\bar{i}_1 + y_1\bar{j}_1 + z_1\bar{k}_1 = x_0\bar{i}_1 + y_0\bar{j}_1 + z_0\bar{k}_1 + x\bar{i} + y\bar{j} + z\bar{k},$$

from where replacing $\bar{i}$, $\bar{j}$ and $\bar{k}$ results in:

$$\begin{Bmatrix} x_1 \\ y_1 \\ z_1 \end{Bmatrix} = \begin{Bmatrix} x_0 \\ y_0 \\ z_0 \end{Bmatrix} + \begin{bmatrix} \cos(x, x_1) \cos(y, x_1) \cos(z, x_1) \\ \cos(x, y_1) \cos(y, x_1) \cos(z, x_1) \\ \cos(x, z_1) \cos(y, x_1) \cos(z, x_1) \end{bmatrix} \begin{bmatrix} x \\ y \\ z \end{bmatrix}, \tag{4.11}$$

or

$$\begin{Bmatrix} x_1 \\ y_1 \\ z_1 \end{Bmatrix} = \begin{Bmatrix} x_0 \\ y_0 \\ z_0 \end{Bmatrix} + [\{i\} \quad \{j\} \quad \{k\}] \begin{Bmatrix} x \\ y \\ z \end{Bmatrix} = \begin{Bmatrix} x_0 \\ y_0 \\ z_0 \end{Bmatrix} + [\mathbf{R}] \begin{Bmatrix} x \\ y \\ z \end{Bmatrix}, \tag{4.12}$$

where it was noted:

$$[\mathbf{R}] = \begin{bmatrix} \cos(x, x_1) \cos(y, x_1) \cos(z, x_1) \\ \cos(x, y_1) \cos(y, x_1) \cos(z, x_1) \\ \cos(x, z_1) \cos(y, x_1) \cos(z, x_1) \end{bmatrix} = [\{i\} \quad \{j\} \quad \{k\}]. \tag{4.13}$$

It is easy to check that $[\mathbf{R}]$ is an orthonormal matrix. Thus,

$$[\mathbf{R}][\mathbf{R}]^T = [\mathbf{R}]^T[\mathbf{R}] = [\mathbf{E}], \tag{4.14}$$

is the matrix relationship that imposes six conditions between the direction cosines. For ease of calculation, we note:

$$\{i\} = \begin{Bmatrix} \alpha_1 \\ \beta_1 \\ \gamma_1 \end{Bmatrix}, \quad \{j\} = \begin{Bmatrix} \alpha_2 \\ \beta_2 \\ \gamma_2 \end{Bmatrix}, \quad \{k\} = \begin{Bmatrix} \alpha_3 \\ \beta_3 \\ \gamma_3 \end{Bmatrix}. \tag{4.15}$$

The following relationships are obtained:

$$\begin{aligned} \alpha_1^2 + \beta_1^2 + \gamma_1^2 = 1; &\quad \alpha_1\alpha_2 + \beta_1\beta_2 + \gamma_1\gamma_2 = 0 \\ \alpha_2^2 + \beta_2^2 + \gamma_2^2 = 1; &\quad \alpha_2\alpha_3 + \beta_2\beta_3 + \gamma_2\gamma_3 = 0\,. \\ \alpha_3^2 + \beta_3^2 + \gamma_3^2 = 1; &\quad \alpha_3\alpha_1 + \beta_3\beta_1 + \gamma_3\gamma_1 = 0 \end{aligned} \tag{4.16}$$

Of the nine direction cosines appearing in the rotation matrix $[\mathbf{R}]$, only three can be chosen independently. In practice, the direction cosines are rarely used as independent parameters. Alternatively, choosing other three parameters, for example the Euler's angles, is easier to use in applications. Direction cosines will be expressed according to these three independent parameters.

Let us now study the problem of eigenvalues and eigenvectors for the rotation matrix [**R**].

The problem of eigenvalues leads to the characteristic equation:

$$([R] - \lambda[E])\{u\} = 0. \tag{4.17}$$

Since the absolute value of a vector does not have to change after a rotation, it follows that the absolute value of the scalar λ must be 1. The condition that the system (4.17) has other solutions besides the null solution is that:

$$\det([R] - \lambda[E]) = 0. \tag{4.18}$$

If Eq. (4.18) is evaluated, the following third-order polynomial equation is obtained:

$$\lambda^3 - I_1\lambda^2 + I_2\lambda - I_3 = 0, \tag{4.19}$$

where the invariants I_i (i = 1, 2, 3) are defined as follows:

$$\begin{aligned} I_1 &= a_{11} + a_{22} + a_{33} = \text{tr}[R], \\ I_2 &= (a_{11}a_{22} - a_{12}a_{21}) + (a_{22}a_{33} - a_{23}a_{32}) + (a_{33}a_{11} - a_{13}a_{31}), \\ I_3 &= \det[R] = 1. \end{aligned} \tag{4.20}$$

The abbreviation tr[***R***] stand for the trace of matrix [***R***].

If we impose the condition $\lambda = 1$ and we take into account that $I_3 = 1$ $I_1 = I_2 = \text{tr}[R]$, then Eq. (4.19) becomes:

$$\lambda^3 - \text{tr}[R]\lambda^2 + \text{tr}[R]\lambda - 1 = 0. \tag{4.21}$$

If we divide the last relation by $\lambda - 1$, it results:

$$\lambda^2 - (\text{tr}[R] - 1)\lambda + 1 = 0, \tag{4.22}$$

with the roots:

$$\lambda_{2,3} = \frac{\text{tr[R]} - 1}{2} \pm \text{i}\sqrt{1 - \left(\frac{\text{tr[R]} - 1}{2}\right)^2} = \cos\varphi \pm \text{i}\sin\varphi = \text{e}^{\pm\text{i}\varphi}, \tag{4.23}$$

where

$$\cos\varphi = \frac{\text{tr[R]} - 1}{2}. \tag{4.24}$$

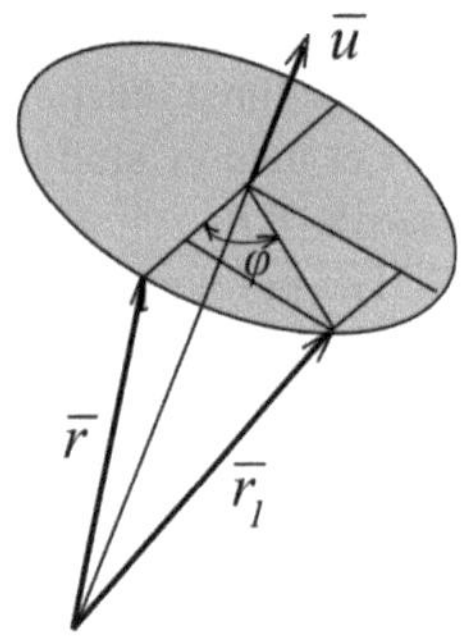

Fig. 4.3 Finite rotation of a rigid body

If $[R]$ is the unit matrix then all three eigenvalues are 1.

If $\text{tr}[R] = -1$ then one eigenvalue is 1 and the other two values are -1.

Consider now a unit vector $\bar{u}$ corresponding to the eigenvalue $\lambda = 1$. This vector results from the Eq. (4.17) if it is considered the normalization relation:

$$u_1^2 + u_2^2 + u_3^2 = 1. \tag{4.25}$$

Equation (4.17) can be written:

$$[R]\{u\} = \lambda\{u\}. \tag{4.26}$$

It results that u_1, u_2, u_3 represent the coordinates of the vector $\bar{u}$ in both coordinate systems, i.e. $Oxyz$ and $O_1x_1y_1z_1$, see Fig. 4.3.

The orientation of the system $Oxyz$ results if a rotation around the vector $\bar{u}$ of the system $O_1x_1y_1z_1$ is performed. A theorem assigned to Euler shows:

The displacement of a body-fixed base (coordinate system $O_1x_1y_1z_1$) from an initial position to an arbitrary final position (coordinate system $Oxyz$) is achieved by a rotation through an angle φ angle about an axis which is fixed in both bases. The axis is determined by the eigenvector associated with the real eigenvalue $\lambda = 1$ of the rotation (direction cosines) matrix $[R]$.

4.1.2 *Euler Angles*

The rotation of the movable coordinate system $Oxyz$ relative to the fixed coordinate system $Ox_1y_1z_1$ is characterized by the nine direction cosines defining the rotation matrix $[\boldsymbol{R}]$. We have seen that these direction cosines are linked by six relationships. It is possible to choose three independent direction cosines to describe the rotation of the rigid body, the other six being able to be expressed according to them. Such a way to study the problem is particularly difficult. There are various ways to choose three independent parameters that describe the finite rotation of the rigid body, the nine direction cosines being expressed according to them.

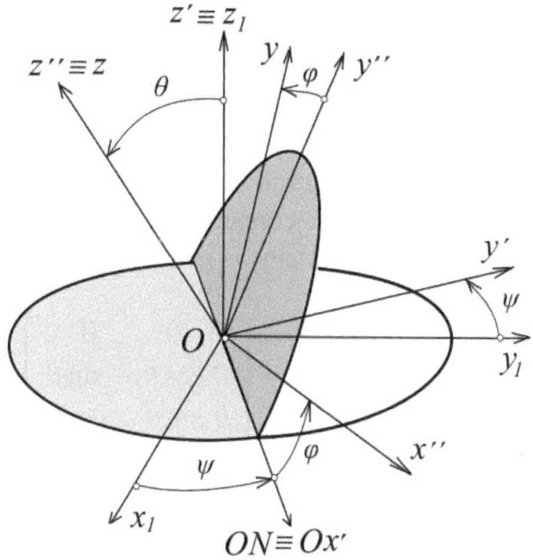

Fig. 4.4 Euler's angles

One way that is especially useful because of its intuitive support is to use Euler's angles as independent parameters. In this way the global rotations are obtained by means of three planar rotations, successively applied to the local reference system. Representing the global and local reference systems, Euler's angles are obtained as follows: intersect the Oxy plane with the $O_1x_1y_1$ plane. The line of intersection is ON and is called the line of nodes. The angle ψ made by Ox_1 with ON is called the precession angle and the angle φ made by Ox with the ON line is called the intrinsic rotation angle. The third parameter that will define the position of the $Oxyz$ reference is the angle θ, called the nutation angle. We will demonstrate in the following how the three previously defined angles can be used as independent parameters for defining the rotation matrix. The angles ψ and φ vary from 0 to 2π (or from $-\pi$ to π), and θ varies between 0 and π (or from $-\pi/2$ to $\pi/2$).

Consider the global coordinate reference system $O_1x_1y_1z_1$ and the coordinate system solidary with the rigid body, $Oxyz$, see Fig. 4.4. The rotation from the global coordinate system to the local coordinate system is done in the following three steps:

- a plane rotation of the coordinate system around the Oz_1-axis with the angle ψ; the coordinate system has a new position defined by the axes $Ox'y'z'$, $Oz \equiv Oz'$. The Ox'-axis is also line of nodes ON;
- $ON = Ox'$ is maintained fixed and a plane angular rotation with the angle θ is performed around it. Thus, the coordinate system has a new position defined by the axes $Ox''y''z''$, $Ox' \equiv Ox''$;
- the Oz-axis is maintained and it rotates around the coordinate system with the angle φ. Doing so, one obtains the coordinate system that defines the final position, $Oxyz$, where $Oz'' \equiv Oz$.

For the first step we have a plane rotation of angle ψ that is defined by the rotation matrix $[R_\psi]$. The relation between the coordinates of a vector expressed in the two reference systems $O_1x_1y_1z_1$ and $Ox'y'z'$ is:

$$\begin{bmatrix} x_1 \\ y_1 \\ z_1 \end{bmatrix} = \begin{bmatrix} \cos\psi & -\sin\psi & 0 \\ \sin\psi & \cos\psi & 0 \\ 0 & 0 & 1 \end{bmatrix} \begin{bmatrix} x' \\ y' \\ z' \end{bmatrix} \text{ or } \{r_1\} = \lfloor R_\psi \rfloor \{r'\}. \tag{4.27}$$

The second rotation of angle θ around the line nodes is expressed by the rotation matrix $[R_\theta]$ and the connection between the components of a vector expressed in the two coordinate systems is given by the relation:

$$\begin{bmatrix} x' \\ y' \\ z' \end{bmatrix} = \begin{bmatrix} 1 & 0 & 0 \\ 0 & \cos\theta & -\sin\theta \\ 0 & \sin\theta & \cos\theta \end{bmatrix} \begin{bmatrix} x'' \\ y'' \\ z'' \end{bmatrix} \text{ or } \{r'\} = [R_\theta]\{r''\}. \tag{4.28}$$

The third rotation of angle φ around the Oz-axis is expressed by the rotation matrix $[R_\varphi]$ and the relation between the components of a vector expressed in the two coordinate systems is given by the relation:

$$\begin{bmatrix} x'' \\ y'' \\ z'' \end{bmatrix} = \begin{bmatrix} \cos\varphi & -\sin\varphi & 0 \\ \sin\varphi & \cos\varphi & 0 \\ 0 & 0 & 1 \end{bmatrix} \begin{bmatrix} x \\ y \\ z \end{bmatrix} \text{ or } \{r''\} = \lfloor R_\varphi \rfloor \{r\}. \tag{4.29}$$

Finally, it can be written:

$$\{r_1\} = \lfloor R_\psi \rfloor \{r'\} = \lfloor R_\psi \rfloor [R_\theta]\{r''\} = \lfloor R_\psi \rfloor [R_\theta] \lfloor R_\varphi \rfloor \{r\}. \tag{4.30}$$

The matrix representing the rotation is:

$$[\mathbf{R}] = \lfloor R_\psi \rfloor [R_\theta] \lfloor R_\varphi \rfloor. \tag{4.31}$$

By doing the calculations, one gets:

$$\begin{aligned}[\mathbf{R}] &= \begin{bmatrix} \cos\psi\cos\varphi - \sin\psi\sin\varphi\cos\theta & -\cos\psi\sin\varphi - \sin\psi\cos\varphi\cos\theta & \sin\psi\sin\theta \\ \sin\psi\cos\varphi + \cos\psi\sin\varphi\cos\theta & -\sin\psi\sin\varphi + \cos\psi\cos\varphi\cos\theta & -\sin\theta\cos\psi \\ \sin\theta\sin\varphi & \sin\theta\cos\varphi & \cos\theta \end{bmatrix} \\ &= [\{i\} \;\; \{j\} \;\; \{k\}]. \end{aligned} \tag{4.32}$$

Thus, the components of the unit vectors $\{i\}$, $\{j\}$, $\{k\}$, defined in the fixed reference system, are obtained. If we consider that the final rotation is obtained by the succession of three angular rotations ψ, θ, φ over a very short time Δt, and if the ratios between angles and time are considered to represent the angular velocity of the rigid body around the three axes, then the angular velocity of the rigid body can be represented as the sum of the three angular velocities:

$$\{\omega\}\dot{\psi}\{k_1\} + \dot{\theta}\{e'\} + \dot{\varphi}\{k\}. \tag{4.33}$$

Taking into account that $\{e'\} = \{i_1\}\cos\psi + \{j_1\}\sin\psi$ and $\{k\}$ as previously defined within the rotation matrix, we have the angular velocity components in the fixed reference system:

$$\begin{aligned}\omega_{x1} &= \dot{\theta}\cos\psi + \dot{\varphi}\sin\theta\sin\psi;\\ \omega_{y1} &= \dot{\theta}\sin\psi - \dot{\varphi}\sin\theta\cos\psi;\\ \omega_{z1} &= \dot{\psi} + \dot{\varphi}\cos\theta.\end{aligned} \tag{4.34}$$

Of course, these components could be obtained by applying the definition relation of the angular speed operator:

$$\{\omega\} = [\dot{R}][R]^T. \tag{4.35}$$

In this case, however, a large number of calculations has to be done. The components of the angular velocity in the local coordinate system are obtained with the relationships

$$\{\Omega\} = [\mathbf{R}]^T\{\omega\}, \tag{4.36}$$

from where the components follow:

$$\begin{aligned}\omega_x &= \dot{\psi}\sin\theta\sin\varphi + \dot{\theta}\cos\varphi;\\ \omega_y &= \dot{\psi}\sin\theta\cos\varphi - \dot{\theta}\sin\varphi;\\ \omega_z &= \dot{\varphi} + \dot{\psi}\cos\theta.\end{aligned} \tag{4.37}$$

4.1.3 Bryan (Cardan) Angles

In this case, the approach is the same as for the Euler rotation, just the order of rotations is different. Consider the global coordinate system $O_1x_1y_1z_1$ and the local coordinate system *Oxyz* (see Fig. 4.5), which is attached to the rigid body. The rotation from the global coordinate system to the local coordinate system is done in the following three steps:

- a plane rotation of the coordinate system around the Ox_1-axis with the angle φ_1; the coordinate system has a new position defined by the axes $Ox'y'z'$, $Ox \equiv Ox'$;
- the Oy'-axis is maintained fixed and performs a plane angular rotation with the angle φ_2 around it; the coordinate system has a new position defined by the axes $Ox''y''Oz''$, $Oy' \equiv Oy''$;
- it is maintained the axis Oz'' and it rotates around the coordinate system with the angle φ_3.

Fig. 4.5 Bryan's angles

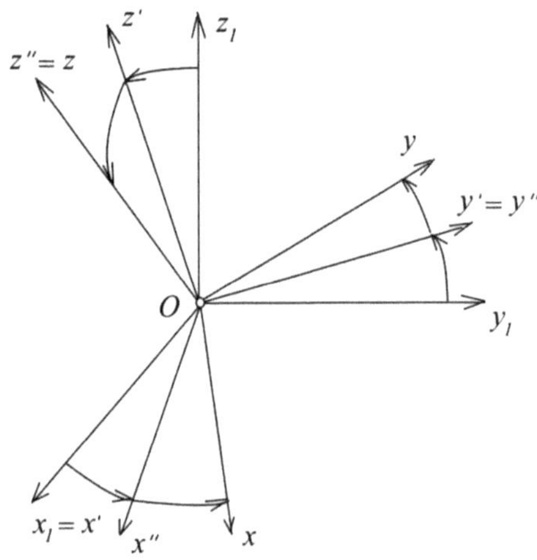

Doing so, one obtains the coordinate system that defines the final position, $Oxyz$, where $Oz'' \equiv Oz$.

The three rotation matrices are in this case:

$$[R_1] = \begin{bmatrix} 1 & 0 & 0 \\ 0 & \cos\varphi_1 & \sin\varphi_1 \\ 0 & -\sin\varphi_1 & \cos\varphi_1 \end{bmatrix} = \begin{bmatrix} 1 & 0 & 0 \\ 0 & c_1 & s_1 \\ 0 & -s_1 & c_1 \end{bmatrix}, \tag{4.38}$$

$$[R_2] = \begin{bmatrix} \cos\varphi_2 & 0 & -\sin\varphi_2 \\ 0 & 1 & 0 \\ \sin\varphi_1 & 0 & \cos\varphi_2 \end{bmatrix} = \begin{bmatrix} c_2 & 0 & -s_2 \\ 0 & 1 & 0 \\ s_1 & 0 & c_2 \end{bmatrix}, \tag{4.39}$$

$$[R_3] = \begin{bmatrix} \cos\varphi_3 & \sin\varphi_3 & 0 \\ -\sin\varphi_3 & \cos\varphi_3 & 0 \\ 0 & 0 & 1 \end{bmatrix} = \begin{bmatrix} c_3 & s_3 & 0 \\ -s_3 & c_3 & 0 \\ 0 & 0 & 1 \end{bmatrix}. \tag{4.40}$$

The rotation matrix is obtained by calculating the product:

$$[R] = [R_1][R_2][R_3]$$

$$[\mathbf{R}] = \begin{bmatrix} c_2c_3 & -c_2s_3 & s_2 \\ c_1s_3 + s_1s_2c_3 & c_1c_3 - s_1s_2s_3 & -s_1c_2 \\ s_1s_3 - c_1s_2c_3 & s_1c_3 + c_1s_2s_3 & c_1c_2 \end{bmatrix}. \tag{4.41}$$

Compared to the representation using Euler's angles, the difference lies in the order in which the plane rotations are made. The advantage of this representation is that it is possible to linearize it in case of small angles. Thus, if $\varphi_i \ll 1 \quad (i = 1, 2, 3)$ the approximations $\sin\varphi_i \cong \varphi_i$ and $\cos\varphi_i \cong 1$ can be made and the rotation matrix becomes:

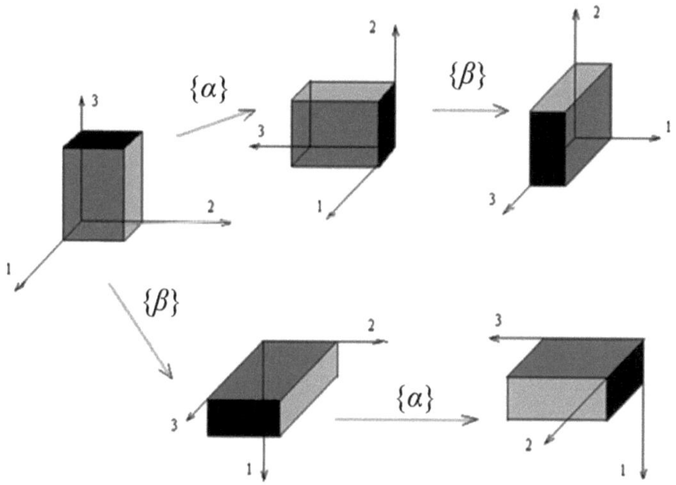

Fig. 4.6 Finite rotations and commutativity

$$[\mathbf{R}]\begin{bmatrix} 1 & -\varphi_3 & \varphi_2 \\ \varphi_3 & 1 & -\varphi_1 \\ -\varphi_2 & \varphi_1 & 1 \end{bmatrix} = \begin{bmatrix} 1\,0\,0 \\ 0\,1\,0 \\ 0\,0\,1 \end{bmatrix} - \begin{bmatrix} 1 & -\varphi_3 & \varphi_2 \\ \varphi_3 & 0 & -\varphi_1 \\ -\varphi_2 & \varphi_1 & 0 \end{bmatrix}$$
$$= [E] - [\varphi]. \tag{4.42}$$

The skew-symmetric matrix allows the representation of a vector product. The components $\varphi_1, \varphi_2, \varphi_3$ are the components of a vector called the finite rotation vector $\bar{\varphi}$. It is interesting to note that the components of this vector in the local coordinate system and the global coordinate system are the same. Indeed:

$$[\mathbf{R}]\{\varphi\} = ([E] - [\varphi])\{\varphi\} = [E]\{\varphi\} - [\varphi]\{\varphi\} = \{\varphi\}. \tag{4.43}$$

4.1.4 Finite Rotations and Commutativity

A counterexample proves that finite rotations are not commutative. Thus, for the sake of simplicity, we designate the axes Ox, Oy, Oz as Axis 1, Axis 2 and Axis 3. Then, if we make two successive rotations of 90°, the first around the Axis 1 and the second around the new Axis 2, one gets the final position of the body presented in Fig. 4.6, the top right position. If we make two rotations of 90° first around the Axis 2 and then around the Axis 1 we get the body placed as in Fig. 4.6, in the bottom right position. The two positions of the body are obviously different.

$$[R_\alpha] = \begin{bmatrix} 1 & 0 & 0 \\ 0 & \cos\frac{\pi}{2} & -\sin\frac{\pi}{2} \\ 0 & \sin\frac{\pi}{2} & \cos\frac{\pi}{2} \end{bmatrix} = \begin{bmatrix} 1 & 0 & 0 \\ 0 & 0 & -1 \\ 0 & 1 & 0 \end{bmatrix};$$

$$[R_\beta] = \begin{bmatrix} \cos\frac{\pi}{2} & 0 & -\sin\frac{\pi}{2} \\ 0 & 1 & 0 \\ \sin\frac{\pi}{2} & 0 & \cos\frac{\pi}{2} \end{bmatrix} = \begin{bmatrix} 0 & 0 & -1 \\ 0 & 1 & 0 \\ 1 & 0 & 0 \end{bmatrix},$$

$$[R_{\alpha+\beta}] = [R_\beta][R_\alpha] = \begin{bmatrix} 0 & 0 & -1 \\ 0 & 1 & 0 \\ 1 & 0 & 0 \end{bmatrix} \begin{bmatrix} 1 & 0 & 0 \\ 0 & 0 & -1 \\ 0 & 1 & 0 \end{bmatrix} = \begin{bmatrix} 0 & -1 & 0 \\ 0 & 0 & -1 \\ 1 & 0 & 0 \end{bmatrix},$$

$$[R_{\beta+\alpha}] = [R_\alpha][R_\beta] = \begin{bmatrix} 1 & 0 & 0 \\ 0 & 0 & -1 \\ 0 & 1 & 0 \end{bmatrix} \begin{bmatrix} 0 & 0 & -1 \\ 0 & 1 & 0 \\ 1 & 0 & 0 \end{bmatrix} = \begin{bmatrix} 0 & 0 & -1 \\ -1 & 0 & 0 \\ 0 & 1 & 0 \end{bmatrix},$$

$$[R_{\alpha+\beta}] \neq [R_{\beta+\alpha}].$$

For the first rotation sequence, the rotation angle is given by:

$$\cos\varphi_1 = \frac{\mathrm{tr}[\mathbf{R}] - 1}{2} = -\frac{1}{2},$$

around the axis having the direction parameters determined by the equations:

$$\begin{bmatrix} -1 & -1 & 0 \\ 0 & -1 & -1 \\ 1 & 0 & -1 \end{bmatrix} \begin{Bmatrix} e_1 \\ e_2 \\ e_3 \end{Bmatrix}_1 = \begin{Bmatrix} 0 \\ 0 \\ 0 \end{Bmatrix}$$

or, if we choose $e_1 = 1$:

$$\begin{matrix} 1 + e_2 = 0 \\ e_2 + e_3 = 0 \end{matrix}.$$

This axis will therefore be determined by the vector:

$$\{u\}_1 = \begin{Bmatrix} e_1 \\ e_2 \\ e_3 \end{Bmatrix}_1 = \begin{Bmatrix} 1 \\ -1 \\ 1 \end{Bmatrix}.$$

For the second rotation sequence, the rotation angle is given by:

$$\cos\varphi_1 = \frac{\mathrm{tr}[\mathbf{R}] - 1}{2} = -\frac{1}{2},$$

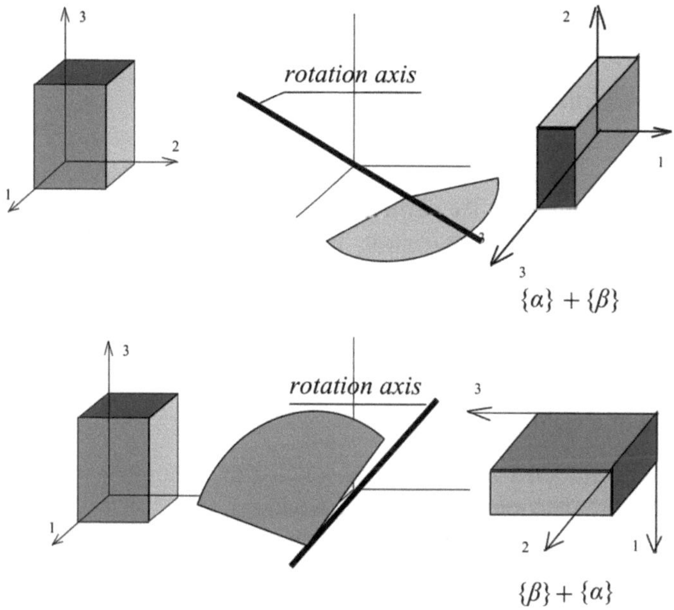

Fig. 4.7 Finite rotation for the example presented in Fig. 4.6

so, just as in the first case, but around the axis having the direction parameters determined by the equations:

$$\begin{bmatrix} -1 & 0 & -1 \\ -1 & -1 & 0 \\ 0 & 1 & -1 \end{bmatrix} \begin{Bmatrix} e_1 \\ e_2 \\ e_3 \end{Bmatrix}_2 = \begin{Bmatrix} 0 \\ 0 \\ 0 \end{Bmatrix}$$

or, if we choose $e_1 = 1$:

$$\begin{aligned} 1 + e_3 &= 0 \\ 1 + e_2 &= 0 \end{aligned}.$$

This axis will therefore be determined by the vector:

$$\{u\}_2 = \begin{Bmatrix} e_1 \\ e_2 \\ e_3 \end{Bmatrix}_2 = \begin{Bmatrix} 1 \\ -1 \\ -1 \end{Bmatrix}.$$

The two vectors around which the finite rotation takes place are therefore different, the angle of rotation being in both cases 120° (Fig. 4.7).

4.2 Moment of Inertia

4.2.1 Fundamental Notions

In the following, the expressions of momentum, angular momentum and kinetic energy will be defined for the particular case of a rigid body, which is a system of material points at which the distance between any two arbitrarily chosen points remains constant. The notion of angular momentum becomes very important for a rigid body, characterized by its rotational motion. The rigidity condition makes the expressions of angular momentum and kinetic energy take particular forms useful in studying the mechanics of a rigid body or a multibody system.

Momentum of a rigid body

It will be considered the rigid body relative to a movable reference system, $Oxyz$. In the presentation the following notations are used to simplify writing: M.P.—material point; S.M.P.—system of material points; R—rigid (regarded as a continuous medium). There are the following momentum expressions:

- for a M.P.:

$$\overrightarrow{H} = m\overrightarrow{v}, \tag{4.44}$$

- for a S.M.P.:

$$\overrightarrow{H} = \sum_{i=1}^{n} m_i \vec{v}_i = \sum_{i=1}^{n} \overrightarrow{H}_i = m\vec{v}_C, \tag{4.45}$$

- for a R.:

$$\overrightarrow{H} = \int d\overrightarrow{H} = \int \vec{v} dm = m\vec{v}_C. \tag{4.46}$$

If the rigid body is considered to be a system of elements with the mass dm, the sum becomes a Riemann integral, so the momentum can be written in the form:

$$\overrightarrow{H} = \int \vec{v} dm = \int (\vec{v}_0 + \overrightarrow{\omega} \times \vec{r}) dm = \vec{v}_0 \int dm + \vec{\omega} \times \int \vec{r} dm = m\vec{v}_0 + \vec{\omega} \times \vec{S}, \tag{4.47}$$

where $\vec{S}$ represents the static moment of the rigid body related to a Cartesian reference system:

$$\vec{S} = \int \vec{r} dm = S_x \vec{i} + S_y \vec{j} + S_z \vec{k}. \tag{4.48}$$

The speed of origin of the mobile reference system is denoted by $\vec{v}_0$ and the angular velocity of the rigid body is $\vec{\omega}$.

If a, b, c are the coordinates of the mass center of the rigid body in the $Oxyz$ system, the components of the static moments are given by:

$$S_x = \int x dm = ma, \quad S_y = \int y dm = mb, \quad S_z = \int z dm = mc, \tag{4.49}$$

from where:

$$\vec{S} = m\vec{r}_C. \tag{4.50}$$

Thus:

$$\vec{H} = m\vec{v}_0 + \vec{\omega} \times m\vec{r}_C = m(\vec{v}_0 + \vec{\omega} \times \vec{r}_C) = m\vec{v}_C,$$

so:

$$\boxed{\vec{H} = m\vec{v}_C.} \tag{4.51}$$

The momentum of a rigid body is equal to the momentum of a material point with mass equal to the mass of the rigid body situated in its center of mass, and participating to the motion of the rigid body.

The relationships (4.4) can also be written as:

$$\{H\} = \begin{Bmatrix} H_x \\ H_y \\ H_z \end{Bmatrix} = m \begin{Bmatrix} v_{0x} \\ v_{0y} \\ v_{0z} \end{Bmatrix} + \begin{bmatrix} 0 & -\omega_z & \omega_y \\ \omega_z & 0 & -\omega_x \\ -\omega_y & \omega_x & 0 \end{bmatrix} \begin{Bmatrix} S_x \\ S_y \\ S_z \end{Bmatrix}$$
$$= m \begin{Bmatrix} v_{0x} \\ v_{0y} \\ v_{0z} \end{Bmatrix} + \begin{Bmatrix} S_z\omega_y - S_y\omega_z \\ S_x\omega_z - S_z\omega_x \\ S_y\omega_x - S_x\omega_y \end{Bmatrix} = m \begin{Bmatrix} v_{0x} \\ v_{0y} \\ v_{0z} \end{Bmatrix} + m \begin{Bmatrix} c\omega_y - b\omega_z \\ a\omega_z - c\omega_x \\ b\omega_x - a\omega_y \end{Bmatrix}, \tag{4.52}$$

or

$$\{H\} = m(\{v_0\} + [\omega]\{r_C\}) = m\{v_C\} = [M]\{v_C\}, \tag{4.53}$$

where

$$[M] = m\begin{bmatrix} 1\,0\,0 \\ 0\,1\,0 \\ 0\,0\,1 \end{bmatrix}; \quad \{\omega\} = \begin{Bmatrix} \omega_x \\ \omega_y \\ \omega_z \end{Bmatrix}; \quad [\omega] = \begin{bmatrix} 0 & -\omega_z & \omega_z \\ \omega_z & 0 & -\omega_x \\ -\omega_y & \omega_x & 0 \end{bmatrix}. \tag{4.54}$$

We also build the matrix of static moments $[S]$ that we will use later:

$$[S] = m\begin{bmatrix} 0 & -S_z & S_y \\ S_z & 0 & -S_x \\ -S_y & S_x & 0 \end{bmatrix}. \tag{4.55}$$

Angular momentum

Starting from the definition, the following expressions can be written for the angular momentum:

- for a P.M.:

$$\vec{K}_O = \vec{r} \times m\vec{v}, \tag{4.56}$$

- for a S.M.P.:

$$\vec{K}_O = \sum_{i=1}^{n} \vec{r}_i \times m_i \vec{v}_i = \sum_{i=1}^{n} \vec{r}_i \times \vec{H}_i, \tag{4.57}$$

- for a rigid body:

$$\vec{K}_O = \int \vec{r} \times \vec{v} dm, \tag{4.58}$$

$$\begin{aligned} \vec{K}_O &= \int \vec{r} \times (\vec{v}_0 + \vec{\omega} \times \vec{r}) dm = \int \vec{r} \times \vec{v}_0 dm + \int \vec{r} \times (\vec{\omega} \times \vec{r}) dm \\ &= \left(\int \vec{r} dm\right) \times \vec{v}_0 + \left(\int r^2 dm\right)\vec{\omega} - \int (\vec{r} \cdot \vec{\omega})\vec{r} dm, \end{aligned} \tag{4.59}$$

where we used the rule of decomposition for a triple vector product. It results:

$$r^2\vec{\omega} = \left(\omega_x \vec{i} + \omega_y \vec{j} + \omega_z \vec{k}\right)\left(x^2 + y^2 + z^2\right); \tag{4.60}$$

$$(\vec{r} \cdot \vec{\omega})\vec{r} = \left(x\omega_x + y\omega_y + z\omega_z\right)\left(x\vec{i} + y\vec{j} + z\vec{k}\right). \tag{4.61}$$

Then:

$$\vec{r} \times (\vec{\omega} \times \vec{r}) = r^2\vec{\omega} - (\vec{r} \cdot \vec{\omega})\vec{r} = \left[\omega_x\left(y^2 + z^2\right) - \omega_y xy - \omega_z xz\right]\vec{i} + \\ +\left[-\omega_x xy + \omega_y\left(z^2 + x^2\right) - \omega_z yz\right]\vec{j} + \left[-\omega_x xz - \omega_y yz + \omega_z\left(x^2 + y^2\right)\right]\vec{k}. \quad (4.62)$$

Using the notations:

$$J_{xx} = \int \left(y^2 + z^2\right)dm; \quad J_{yy} = \int \left(z^2 + x^2\right)dm; \quad J_{zz} = \int \left(x^2 + y^2\right)dm; \\ J_{xy} = \int xydm; \quad J_{yz} = \int yzdm; \quad J_{zx} = \int zxdm, \quad (4.63)$$

one obtains:

$$\int \vec{r} \times (\vec{\omega} \times \vec{r})dm = \left[\omega_x J_{xx} - \omega_y J_{xy} - \omega_z J_{xz}\right]\vec{i} \\ +\left[-\omega_x J_{xz} + \omega_y J_{yy} - \omega_z J_{yz}\right]\vec{j} + \left[-\omega_x J_{xx} - \omega_y J_{yx} + \omega_z J_{zz}\right]\vec{k}. \quad (4.64)$$

Thus,

$$\vec{K}_O = \vec{S} \times \vec{v}_0 + \left(J_{xx}\omega_x - J_{xy}\omega_y - J_{xz}\omega_z\right)\vec{i} + \left(J_{yy}\omega_y - J_{yz}\omega_z - J_{yx}\omega_x\right)\vec{j} \\ +\left(J_{zz}\omega_z - J_{yz}\omega_y - J_{xz}\omega_x\right)\vec{k}, \quad (4.65)$$

or in matrix form:

$$\{K_O\} = [S]\{v_0\} + [J_O]\{\omega\}, \quad (4.66)$$

where:

$$[J_O] = \begin{bmatrix} J_{xx} & -J_{xy} & -J_{xz} \\ -J_{yx} & J_{yy} & -J_{yz} \\ -J_{zx} & -J_{zy} & J_{zz} \end{bmatrix} \text{ and } \{K_O\} = \begin{Bmatrix} K_{Ox} \\ K_{Oy} \\ K_{Oz} \end{Bmatrix}. \quad (4.67)$$

The matrix $[J_O]$ is called the matrix of the moments of inertia. The matrix components J_{xx}, J_{yy}, J_{zz} are called the moments of inertia, and J_{xy}, J_{yz}, J_{zx} the products of inertia. The components of the kinetic moment vector are:

$$K_{Ox} = mbv_{0z} - mcv_{0y} + J_{xx}\omega_x - J_{xy}\omega_y - J_{xz}\omega_z, \\ K_{Oy} = mcv_{0x} - mav_{0z} + J_{yy}\omega_y - J_{yz}\omega_z - J_{zx}\omega_x, \\ K_{Oz} = mav_{0y} - mbv_{0x} + J_{zz}\omega_z - J_{zx}\omega_x - J_{xy}\omega_y. \quad (4.68)$$

If we use the matrix expression:

$$[r] = \begin{bmatrix} 0 & -z & y \\ z & 0 & -x \\ -y & z & 0 \end{bmatrix}, \tag{4.69}$$

then it results for the matrix of moments of inertia the alternative expression:

$$\boxed{[J_O] = -\int [r]\,[r]\,\mathrm{d}m}, \tag{4.70}$$

which is useful to simplify some applicatios.

If $O \equiv C$ (the center of the mass of the rigid body), then the angular moment has the expression:

$$\boxed{\{K_C\} = [J_C]\{\omega\}.} \tag{4.71}$$

Kinetic energy of a rigid body

Starting for the definition, the following expressions can be obtained:

- for a P.M.:

$$E_c = \frac{1}{2}mv^2, \tag{4.72}$$

- for a S.M.P.:

$$E_c = \sum_{i=1}^{n} \frac{1}{2}m_i v_i^2, \tag{4.73}$$

- for a rigid body:

$$E_c = \frac{1}{2}\int v^2 dm. \tag{4.74}$$

It results:

$$E_c = \frac{1}{2}\int (\bar{v}_0 + \bar{\omega} \times \bar{r})^2 dm = \frac{1}{2}v_0^2 \int dm + \int \bar{v}_0(\bar{\omega} \times \bar{r})dm + \frac{1}{2}\int (\bar{\omega} \times \bar{r})^2 dm. \tag{4.75}$$

From Lagrange's identity

$$(\bar{\omega} \times \bar{r})^2 + (\bar{\omega}\bar{r})^2 = \omega^2 r^2, \tag{4.76}$$

(the demonstration is immediate) one obtains:

$$(\bar{\omega} \times \bar{r})^2 = \omega^2 r^2 - (\bar{\omega}\bar{r})^2.$$

Thus:

$$\begin{aligned}(\bar{\omega} \times \bar{r})^2 &= \left(\omega_x^2 + \omega_y^2 + \omega_z^2\right)\left(x^2 + y^2 + z^2\right) - \left(x\omega_x + y\omega_y + z\omega_z\right)^2 \\ &= \left(y^2 + z^2\right)\omega_x^2 + \left(z^2 + x^2\right)\omega_y^2 + \left(x^2 + y^2\right)\omega_z^2 \\ &\quad - 2xy\omega_x\omega_y - 2yz\omega_y\omega_z - 2zx\omega_z\omega_x. \end{aligned} \tag{4.77}$$

It results

$$\begin{aligned}&\int (\bar{\omega} \times \bar{r})^2 dm \\ &= J_{xx}\omega_x^2 + J_{yy}\omega_y^2 + J_{zz}\omega_z^2 - 2J_{xy}\omega_x\omega_y - 2J_{yz}\omega_y\omega_z - 2J_{zx}\omega_z\omega_x. \end{aligned} \tag{4.78}$$

We can write:

$$\int (\bar{\omega} \times \bar{r})^2 dm = \{\omega\}^T [J_C]\{\omega\}. \tag{4.79}$$

Then:

$$\begin{aligned} E_c &= \frac{1}{2}\left(mv_0^2 + \{\omega\}^T[S]\{v_0\} + \{v_0\}^T[S]\{\omega\} + \{\omega\}^T[J]\{\omega\}\right) \\ &= \frac{1}{2}\left[\{v_0\}^T \{\omega\}^T\right]\begin{bmatrix} [M] & [S]^T \\ [S] & [J] \end{bmatrix}\begin{Bmatrix} \{v_0\} \\ \{\omega\} \end{Bmatrix}. \end{aligned} \tag{4.80}$$

If we are in the center of mass we have $[S] = 0$, and then:

$$\boxed{E_c = \tfrac{1}{2}mv^2 + \tfrac{1}{2}\{\omega\}^T[J_C]\{\omega\}.} \tag{4.81}$$

4.2.2 Moment of Inertia; Definitions

Let us consider a system of n material points with mass m_i, $i = 1, \ldots, n$ and denote by δ_i the distance of the point A_i to a straight line (Δ_i). The moment of inertia (or angular mass or rotational inertia) of the system of material points about the straight line (Δ_i) is the sum J_Δ given by:

$$J_\Delta = \sum_{i=1}^{n} m_i \delta_i^2. \tag{4.82}$$

For a rigid composite system the moment of inertia is the sum of the moments of inertia of its subsystems (considered about the same axis).

Generally, if the system of points is related to an orthogonal axis system $Oxyz$, the expressions of the form $\sum m_i x_i^r y_i^p z_i^q$ where r, p, q are positive integers are called moments. The sum $o = r + p + q$ is called the order of the moment.

In mechanics, the following types of moments are used:

(a) moment of order zero:

$$\sum_{i=1}^{n} m_i = m, \tag{4.83}$$

which represents the total mass of the body.

(b) first order moments:

$$S_x = \sum m_i x_i, \quad S_x = \sum m_i y_i, \quad S_x = \sum m_i z_i, \tag{4.84}$$

which are called the static moments in mechanics (defined in 4.1.1).

(c) second order moments are:

- Axial moment of inertia:

$$J_\Delta = \sum_i m_i \delta_i^2, \tag{4.85}$$

where δ_i^2 represents the distance from the point m_i to the line. If the axes of the coordinate system are considered, we have the moments of inertia with respect to these axes defined in the following way:

$$\begin{aligned} J_{xx} &= \sum_i m_i \left(y_i^2 + z_i^2\right) = \sum_i m_i \delta_{ix}^2, \\ J_{yy} &= \sum_i m_i \left(z_i^2 + x_i^2\right) = \sum_i m_i \delta_{iy}^2, \\ J_{zz} &= \sum_i m_i \left(x_i^2 + y_i^2\right) = \sum_i m_i \delta_{iz}^2. \end{aligned} \tag{4.86}$$

where the quantities δ_{ix}, δ_{iy}, δ_{iz} represent the distances from the points to the corresponding axes of the coordinate system.

- Product of inertia:

$$J_{xy} = \sum_i m_i x_i y_i, \quad J_{yz} = \sum_i m_i y_i z_i, \quad J_{zx} = \sum_i m_i z_i x_i. \tag{4.87}$$

- Planar moment of inertia:

$$J_\pi = \sum_i m_i \delta_i^2, \tag{4.88}$$

where δ_i^2 is the distance from the mass point m_i to the plane π. If a Cartesian coordinate system is considered, we have:

$$\begin{aligned} J_{xOy} &= \sum_i m_i z_i^2 = \sum_i m_i \delta_{iz}^2, \\ J_{yOz} &= \sum_i m_i x_i^2 = \sum_i m_i \delta_{ix}^2, \\ J_{zOx} &= \sum_i m_i y_i^2 = \sum_i m_i \delta_{iy}^2, \end{aligned} \tag{4.89}$$

where the quantities $\delta_{ix}, \delta_{iy}, \delta_{iz}$ represents the distance to the planes of the coordinate system.

- Polar moment of inertia:

$$J_O = \sum_i m_i \delta_i^2 = \sum_i m_i \left(x_i^2 + y_i^2 + z_i^2\right), \tag{4.90}$$

where δ_i represents the distance from point i to the origin of the coordinates system.

In the case of a rigid body with the mass continuously distributed, the sums will become Riemann sums and the expressions of the moment of inertia will be:

- Axial moments of inertia:

$$J_{xx} = \int \left(y^2 + z^2\right) \mathrm{d}m, \quad J_{yy} = \int \left(z^2 + x^2\right) \mathrm{d}m, \quad J_{zz} = \int \left(x^2 + y^2\right) \mathrm{d}m. \tag{4.91}$$

- Product of inertia:

$$J_{xy} = \int xy \, \mathrm{d}m, \quad J_{yz} = \int yz \, \mathrm{d}m, \quad J_{zx} = \int zx \, \mathrm{d}m. \tag{4.92}$$

- Planar moments of inertia:

$$J_{xOy} = \int z^2 \, \mathrm{d}m, \, J_{yOz} = \int x^2 \, \mathrm{d}m, \quad J_{zOx} = \int y^2 \mathrm{d}m. \tag{4.93}$$

- Polar moment of inertia:

$$J_O = \int \left(x^2 + y^2 + z^2\right) \mathrm{d}m. \tag{4.94}$$

Integrals will be evaluated over the entire domain defined by the rigid body. Moments of inertia characterize the inertia of the rigid body in rotation. Dimensional, the inertia moment J is expressed by: $\langle J \rangle = ML^2$. The moment of the inertia matrix plays, for the general rotation of the rigid, the same role that the mass matrix plays for the translational motion.

4.2.3 Rotation of the Coordinates System

The study of the rotation matrix was exposed in Sect. 4.1.1. We review some of the useful notions. Consider the reference system $Ox_1y_1z_1$, of the same origin as the $Oxyz$ system, but with other directions of the axes, defined by the unit vectors (see Fig. 4.2):

$$\{i\} = \begin{Bmatrix} \cos(x, x_1) \\ \cos(x, y_1) \\ \cos(x, z_1) \end{Bmatrix} = \begin{Bmatrix} \alpha_1 \\ \beta_1 \\ \gamma_1 \end{Bmatrix},$$

$$\{j\} = \begin{Bmatrix} \cos(y, x_1) \\ \cos(y, y_1) \\ \cos(y, z_1) \end{Bmatrix} = \begin{Bmatrix} \alpha_2 \\ \beta_2 \\ \gamma_2 \end{Bmatrix},$$

$$\{k\} = \begin{Bmatrix} \cos(z, x_1) \\ \cos(z, y_1) \\ \cos(z, z_1) \end{Bmatrix} = \begin{Bmatrix} \alpha_3 \\ \beta_3 \\ \gamma_3 \end{Bmatrix}. \tag{4.95}$$

The matrix

$$[R] = [\{i\} \quad \{j\} \quad \{k\}] = \begin{bmatrix} \alpha_1 & \alpha_2 & \alpha_3 \\ \beta_1 & \beta_2 & \beta_3 \\ \gamma_1 & \gamma_2 & \gamma_3 \end{bmatrix},$$

represents the rotation matrix whose components are the direction cosines of the unit vectors defining the new axes.

A vector $\{\mathrm{a}\}$ with the components expressed in the $Oxyz$ system will become the vector $\{\mathrm{a}'\}$ with components expressed in the $Ox_1y_1z_1$ system:

$$\{a'\} = a_1\{i\} + a_2\{j\} + a_3\{k\} = [\{i\} \quad \{j\} \quad \{k\}]\begin{Bmatrix} a_1 \\ a_2 \\ a_3 \end{Bmatrix} = [R]\{a\}. \tag{4.96}$$

The matrix $[R]$ has the property:

$$[R]^T[R] = [E], \tag{4.97}$$

where $[E]$ is the unit matrix, a relationship that will be demonstrated below (matrix $[R]$ is orthonormal). The unit vectors $\{i\}$, $\{j\}$, $\{k\}$ are orthonormals, so:

$$\{i\}^T\{i\} = \{j\}^T\{j\} = \{k\}^T\{k\} = 1; \quad \{i\}^T\{j\} = \{j\}^T\{k\} = \{k\}^T\{i\} = 0. \tag{4.98}$$

Then:

$$\begin{aligned} [R]^T[R] &= \begin{bmatrix} \{i\}^T \\ \{j\}^T \\ \{k\}^T \end{bmatrix}\left[\{i\}^T \quad \{j\}^T\{k\}^T\right] \\ &= \begin{bmatrix} \{i\}^T\{i\} & \{i\}^T\{j\} & \{i\}^T\{k\} \\ \{j\}^T\{i\} & \{j\}^T\{j\} & \{j\}^T\{k\} \\ \{k\}^T\{i\} & \{k\}^T\{j\} & \{k\}^T\{k\} \end{bmatrix} = \begin{bmatrix} 1 & 0 & 0 \\ 0 & 1 & 0 \\ 0 & 0 & 1 \end{bmatrix} = [E]. \end{aligned} \tag{4.99}$$

It then follows: $[R]^{-1} = [R]^T$ and thus:

$$\{a\} = [R]^T\{a'\}. \tag{4.100}$$

To determine how the matrix of moments of inertia changes after a rotation of the coordinate system, an easy way is to see how the angular moment vector changes through the rotation of the coordinate system.

The angular moment in a coordinate system $Oxyz$, whose origin moves with the velocity $\{v_0\}$ and has the angular velocity $\{\omega\}$ is:

$$\{K_O\} = [S]\{v_0\} + [J_O]\{\omega\}. \tag{4.101}$$

In the new coordinates system the kinetic momentum becomes

$$\begin{aligned} \{K'_O\} &= [R]^T\{K_O\} = [R]^T[S]\{v_0\} + [R]^T[J_O]\{\omega\} \\ &= [R]^T[S][R]\{v'_0\} + [R]^T[J_O][R][R]^T\{\omega\} \\ &= [R]^T[S][R]\{v'_0\} + [R]^T[J_O][R]\{\omega'\}. \end{aligned} \tag{4.102}$$

But we must have the relationship:

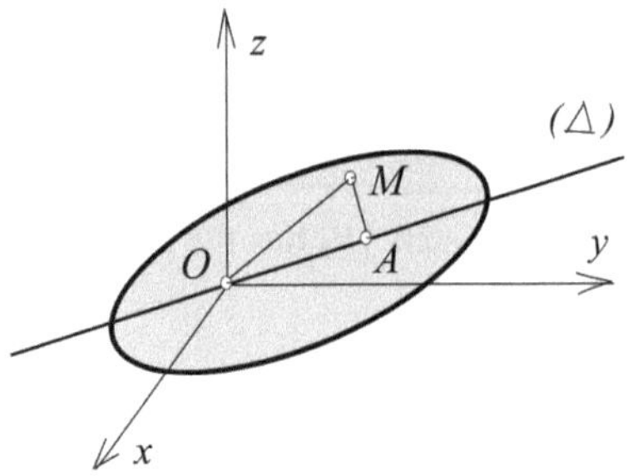

Fig. 4.8 Moment of inertia around an axis

$$\{K'_O\} = [S']\{v'_0\} + [J'_O]^T\{\omega'\}, \tag{4.103}$$

from where by identifying with (4.53) it results:

$$[J'_O] = [R]^T[J_O][R], \tag{4.104}$$
$$[S'] = [R]^T[S][R]. \tag{4.105}$$

The written relationship indicates how the matrix of moments of inertia changes after a rotation of axes. From Eq. (4.104), it is possible to obtain $[J_O]$ as:

$$[J_O] = [R][J'_O][R]^T, \tag{4.106}$$

which is easy to obtain by calculus.

We exemplify the calculation of the moment of inertia to the Ox'-axis' if the matrix of the moments of inertia $[J_O]$ is known:

$$\begin{aligned} J'_{xx} &= \{i\}^T[J_O]\{i\} \\ &= J_{xx}\alpha_1^2 + J_{yy}\beta_1^2 + J_{zz}\gamma_1^2 - 2J_{xy}\alpha_1\beta_1 - 2J_{yz}\beta_1\gamma_1 - 2J_{zz}\gamma_1\alpha_1 \end{aligned}. \tag{4.107}$$

4.2.4 Moment of Inertia of a Body Around an Axis

If a rigid body is referred to a coordinate system $Oxyz$, the axial moments are been defined as (Fig. 4.8):

$$\begin{aligned} J_{xx} &= \int (y^2 + z^2)\,\mathrm{d}m = \int \delta_x^2\,\mathrm{d}m; \\ J_{yy} &= \int (z^2 + x^2)\,\mathrm{d}m = \int \delta_y^2\,\mathrm{d}m; \\ J_z &= \int (x^2 + y^2)\,\mathrm{d}m = \int \delta_z^2\,\mathrm{d}m. \end{aligned} \tag{4.108}$$

where the distances δ_x, δ_y, δ_z represent the distances of the current point to the axes Ox, respectively Oy and Oz.

In the following, it is expressed the moment around an axis using the values of the axial moments and the products of inertia. The moment around an axis (Δ) was defined by the relation:

$$J_\Delta = \int \delta^2 \, \mathrm{d}m, \tag{4.109}$$

where δ represents the distance from a point to the axis (Δ). If $\vec{u}(\alpha, \beta, \gamma)$ it is the unit vector of the axis, it follows:

$$\delta^2 = r^2 - (\vec{r}\vec{u})^2 = r^2u^2 - (\vec{r}\vec{u})^2 = \alpha^2(y^2 + z^2) + \beta^2(z^2 + x^2) + \gamma^2(x^2 + y^2)$$
$$- 2\alpha\beta xy - 2\beta\gamma yz - 2\gamma\alpha xz = \begin{bmatrix} \alpha & \beta & \gamma \end{bmatrix} \begin{bmatrix} y^2 + z^2 & -xy & -xz \\ -yx & z^2 + x^2 & -yz \\ -zx & -zy & x^2 + y^2 \end{bmatrix} \begin{Bmatrix} \alpha \\ \beta \\ \gamma \end{Bmatrix}. \tag{4.110}$$

Thus:

$$J_\Delta = \{u\}^T [J_O] \{u\}, \tag{4.111}$$

that is, it is possible to express the moment about an axis using a quadratic form defined by the matrix $[J_O]$. For example, it can be written:

$$J_{xx} = \{i\}^T [J_o] \{i\}, \quad J_{xy} = \{i\}^T [J_o] \{j\}, \quad \text{etc.}, \tag{4.112}$$

where: $\{i\} = \begin{bmatrix} 1 \, 0 \, 0 \end{bmatrix}^T$, $\{j\} = \begin{bmatrix} 0 \, 1 \, 0 \end{bmatrix}^T$.

4.2.5 Directions of Extremum for the Moments of Inertia

The moment of inertia around an axis takes different values for different directions. It may be a question of looking for an axis for which J_Δ admits an extremal value. The Lagrange multipliers method will be used. Direction cosines (angles) α, β, γ are subject to the following restrictions:

$$\alpha^2 + \beta^2 + \gamma^2 = 1. \tag{4.113}$$

Thus:

$$g(\alpha, \beta, \gamma) = 1 - (\alpha^2 + \beta^2 + \gamma^2) = 0. \tag{4.114}$$

The Lagrangean associated with the problem is:

$$f(\alpha, \beta, \gamma) = J_\Delta + \lambda g(\alpha, \beta, \gamma)$$
$$= \begin{bmatrix} \alpha \; \beta \; \gamma \end{bmatrix} [J_O] \begin{Bmatrix} \alpha \\ \beta \\ \gamma \end{Bmatrix} + \lambda\left(1 - \alpha^2 - \beta^2 - \gamma^2\right). \tag{4.115}$$

We have used the following result for the derivation: if $[A]$ is a symmetric matrix and

$$b = \frac{1}{2}\{X\}^T [A]\{X\}, \tag{4.116}$$

then we have:

$$\begin{Bmatrix} \frac{db}{dX} \end{Bmatrix} = \begin{bmatrix} \frac{\partial b}{\partial x_1} \frac{\partial b}{\partial x_2} \cdots \frac{\partial b}{\partial x_n} \end{bmatrix} = [A]\{X\}. \tag{4.117}$$

The conditions (4.73) can also be written as:

$$[J_O] \begin{Bmatrix} \alpha \\ \beta \\ \gamma \end{Bmatrix} = \lambda \begin{Bmatrix} \alpha \\ \beta \\ \gamma \end{Bmatrix}, \tag{4.118}$$

which expresses the fact that the sought direction $\vec{u}(\alpha, \beta, \gamma)$ is an eigenvector for the matrix $[J_O]$, and the values λ are eigenvalues for $[J_O]$. Therefore, the extreme values of the moments of inertia around an axis are the eigenvalues of the matrix $[J_O]$, and the directions for which this happens are the eigenvectors of $[J_O]$.

Their eigenvalues are also called the principal moments of inertia, and their eigenvectors, the principal directions of inertia. This relationship can also be written as:

$$[J_O]\{u\} = \lambda\{u\}, \tag{4.119}$$

or

$$[J_O]\{u\} = \lambda[E]\{u\},$$

or still:

$$([J_O] - \lambda[E])\{u\} = 0, \tag{4.120}$$

which is a homogeneous linear system that has a different solution than the ordinary one if and only if:

$$\det([J_O] - \lambda[E]) = 0. \tag{4.121}$$

This relationship is an equation of third degree in λ, called the characteristic equation, which provides the eigenvalues. The characteristic equation can be represented in the following form:

$$\lambda^3 - I_1\lambda^2 + I_2\lambda - I_3 = 0, \tag{4.122}$$

where I_1, I_2, I_3 represent the invariants of the matrix of moments of inertia (in every reference system in which they are calculated, if they have the same origin, the sizes will have the same value) and are given by:

$$I_1 = J_1 + J_2 + J_3 = J_{xx} + J_{yy} + J_{zz},$$

$$I_2 = J_1J_2 + J_2J_3 + J_3J_1 = \begin{vmatrix} J_{xx} & -J_{xy} \\ -J_{xy} & J_{yy} \end{vmatrix} + \begin{vmatrix} J_{yy} & -J_{yz} \\ -J_{yz} & J_{zz} \end{vmatrix} + \begin{vmatrix} J_{xx} & -J_{xz} \\ -J_{xz} & J_{zz} \end{vmatrix},$$

$$I_3 = J_1J_2J_3 = \begin{vmatrix} J_{xx} & -J_{xy} & -J_{xz} \\ -J_{xy} & J_{yy} & -J_{yz} \\ -J_{xz} & -J_{yz} & J_{zz} \end{vmatrix},$$

where J_1, J_2, J_3 are solutions of Eq. (4.122). If the values J_1, J_2, J_3 are determined, returning to the homogeneous system (4.120) we can determine the eigenvectors.

4.2.6 A Property of the Principal Direction of Inertia

The matrix of the moments of inertia is a real, symmetric, positively defined matrix. This allows the determination of characteristic properties, which facilitate the study of such problems. The main property that characterizes this matrix is:

The principal directions of inertia are orthogonal. Thus, if we note with $\{e_1\}$, $\{e_2\}$, $\{e_3\}$ system solutions:

$$([J_O] - \lambda_i[E])\{e_i\} = 0, \tag{4.123}$$

for $i = 1, 2, 3$ then $\{e_i\}$, $i = 1, 2, 3$ give the principal directions of inertia. It is noted with J_1, J_2, J_3 its eigenvalues, so the principal moments of inertia. It is obtained:

$$[J_O]\{e_i\} = J_i\{e_i\},$$

which by pre-multiplying with $\{e_j\}^{\mathrm{T}}$ becomes:

$$\{e_j\}^T[J_O]\{e_i\} = J_i\{e_j\}^T\{e_i\}.$$

In the same way:

$$\{e_i\}^T[J_O]\{e_j\} = J_j\{e_i\}^T\{e_j\}.$$

If $i = j$ then it results that:

$$\{e_i\}^T[J_O]\{e_i\} = J_i, \tag{4.124}$$

whereas $\{e_i\}^T\{e_i\} = 1$. If $i \neq j$, subtracting the two relationships, we obtain:

$$0 = (J_i - J_j)\{e_i\}^T\{e_j\}, \tag{4.125}$$

from where, since $J_i \neq J_j$, it generally results that $\{e_i\}^T\{e_j\} = 0$, i.e. the two principal directions are orthogonal. It was taken into account that:

$$\{e_i\}^T[J_O]\{e_j\} = \{e_j\}^T[J_O]\{e_i\}, \tag{4.126}$$

because $[J_O]$ is a symmetrical matrix and:

$$\{e_i\}^T\{e_j\} = \{e_j\}^T\{e_i\}, \tag{4.127}$$

because the scalar product (dot) is commutative. Thus, the unit vectors $\{e_1\}$, $\{e_2\}$, $\{e_3\}$ will form an orthogonal coordinate system.

The matrix:

$$[R] = \left[\ \{e_1\}\ \{e_2\}\ \{e_3\}\ \right],$$

makes the transition from the *Oxyz* coordinate system defined by the principal directions of inertia.

Taking into account the previously demonstrated relationship, written as:

$$\{e_j\}^T[J_O]\{e_i\} = J_i\delta_{ij}, \tag{4.128}$$

where δ_{ij} is the Kronecker's symbol, it follows:

$$[R]^T[J_O][R] = \begin{bmatrix} \{e_1\}^T \\ \{e_2\}^T \\ \{e_3\}^T \end{bmatrix} [J_O][\{e_1\}\ \{e_2\}\ \{e_3\}]$$
$$= \begin{bmatrix} \{e_1\}^T[J_O]\{e_1\} & \{e_1\}^T[J_O]\{e_2\} & \{e_1\}^T[J_O]\{e_3\} \\ \{e_2\}^T[J_O]\{e_1\} & \{e_2\}^T[J_O]\{e_2\} & \{e_2\}^T[J_O]\{e_3\} \\ \{e_3\}^T[J_O]\{e_1\} & \{e_3\}^T[J_O]\{e_2\} & \{e_3\}^T[J_O]\{e_3\} \end{bmatrix} = \begin{bmatrix} J_1 & 0 & 0 \\ 0 & J_2 & 0 \\ 0 & 0 & J_3 \end{bmatrix}, \tag{4.129}$$

Fig. 4.9 Inertia ellipsoid

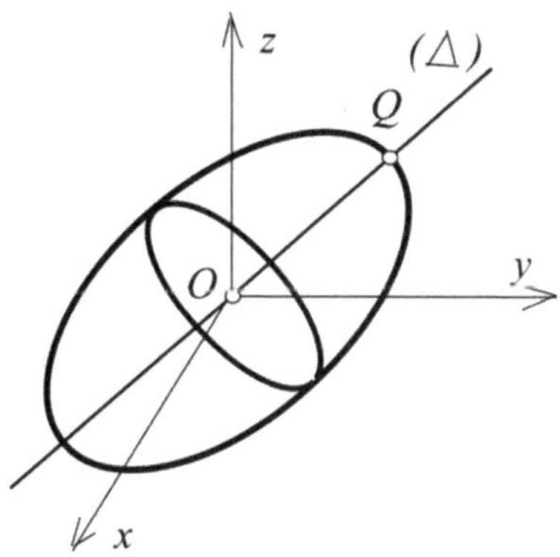

that is, to this coordinate system, the matrix of the moments of inertia has a diagonal form. This property is true for any real symmetric matrix and is called the eigen-decomposition (Sylvester's law of inertia). It is therefore convenient to choose the principal coordinate system for dynamic problems, as the terms that will appear in the equations will be fewer.

4.2.7 Inertia Ellipsoid

The inertia ellipsoid (or Poinsot's ellipsoid) is an intuitive representation of the rotating inertial properties of the bodies. Take on the axis (Δ) whose direction is given by the unit vector $\{u\} = \begin{bmatrix} \alpha \; \beta \; \gamma \end{bmatrix}^T$ an OQ segment inversely proportional to the square root of J_Δ:

$$OQ = \frac{k}{\sqrt{J_\Delta}}, \tag{4.130}$$

(where k is an arbitrary constant, whose dimensions are chosen so that OQ results in a length). Thus:

$$\{r\} = OQ = \frac{k}{\sqrt{J_\Delta}}\{u\}. \tag{4.131}$$

The coordinates of the point Q are:

$$x = \frac{k}{\sqrt{J_\Delta}}\alpha; \quad y = \frac{k}{\sqrt{J_\Delta}}\beta; \quad z = \frac{k}{\sqrt{J_\Delta}}\gamma. \tag{4.132}$$

Thus:

$$\begin{Bmatrix} x \\ y \\ z \end{Bmatrix} = \frac{k}{\sqrt{J_\Delta}} \begin{Bmatrix} \alpha \\ \beta \\ \gamma \end{Bmatrix}, \tag{4.133}$$

from where:

$$\{u\} = \begin{Bmatrix} \alpha \\ \beta \\ \gamma \end{Bmatrix} = \frac{\sqrt{J_\Delta}}{k} \begin{Bmatrix} x \\ y \\ z \end{Bmatrix}. \tag{4.134}$$

Assuming that the direction of the line (Δ) varies, always passing through O, the geometric point of Q is an ellipsoid. Indeed:

$$J_\Delta = \{u\}^T [J_O] \{u\} = \frac{J_\Delta}{k^2} \begin{bmatrix} x \; y \; z \end{bmatrix} [J_O] \begin{Bmatrix} x \\ y \\ z \end{Bmatrix} \tag{4.135}$$

leads to

$$\begin{bmatrix} x \; y \; z \end{bmatrix} [J_O] \begin{Bmatrix} x \\ y \\ z \end{Bmatrix} = k^2, \tag{4.136}$$

which is the equation of a quadric with center in O (Fig. 4.9). The quadric has no points at infinity as:

$$\begin{Bmatrix} x \\ y \\ z \end{Bmatrix} = \frac{1}{\sqrt{J_\Delta}} \begin{Bmatrix} \alpha \\ \beta \\ \gamma \end{Bmatrix}. \tag{4.137}$$

$J_\Delta \neq 0$ (except if all points are on the straight line Δ), and α, β, γ are finite because they are direction cosines. Thus, it is an ellipsoid, called the inertia ellipsoid, corresponding to point O.

In the global coordinate system, if X, Y, Z are the coordinates of the current point in this system, the equation of the ellipsoid will take the form:

$$\begin{bmatrix} X \; Y \; Z \end{bmatrix} \begin{bmatrix} J_1 & 0 & 0 \\ 0 & J_2 & 0 \\ 0 & 0 & J_3 \end{bmatrix} \begin{Bmatrix} X \\ Y \\ Z \end{Bmatrix} = k^2, \tag{4.138}$$

or

$$J_1 X^2 + J_2 Y^2 + J_3 Z^2 = k^2. \tag{4.139}$$

The relationship obtained from (4.136) can be written in the form:

$$\begin{bmatrix} x & y & z \end{bmatrix}[R][R]^T[J_O][R][R]^T \begin{Bmatrix} x \\ y \\ z \end{Bmatrix} = k^2, \tag{4.140}$$

and taking into account the fact that:

$$\begin{Bmatrix} X \\ Y \\ Z \end{Bmatrix} = [R]^T \begin{Bmatrix} x \\ y \\ z \end{Bmatrix} \quad \text{and} \quad \begin{bmatrix} X & Y & Z \end{bmatrix} = \begin{bmatrix} x & y & z \end{bmatrix}[R].$$

The principal directions of inertia are the directions of the semi-principal diameter of the ellipsoid. Related to the principal directions of inertia, the inertia products are null. This property sometimes allows the identification of the principal axes of inertia. For example, if a body has an axis of symmetry, then the inertial product in a plane determined by the axis and a perpendicular axis are null and the axis is the principal axis of inertia for all the points through which it passes.

The equation of the inertia ellipsoid relative to the principal direction coordinate system can also be written as:

$$\frac{X^2}{\left(\sqrt{\frac{1}{J_1}}\right)^2} + \frac{Y^2}{\left(\sqrt{\frac{1}{J_2}}\right)^2} + \frac{Z^2}{\left(\sqrt{\frac{1}{J_3}}\right)^2} = k^2, \tag{4.141}$$

where the half-axes (semi-principal diameters) a, b, c are:

$$a = \frac{k}{\sqrt{J_1}}, \quad b = \frac{k}{\sqrt{J_2}}, \quad c = \frac{k}{\sqrt{J_3}}, \tag{4.142}$$

and the ellipsoid equation has the classical form:

$$\frac{X^2}{a^2} + \frac{Y^2}{b^2} + \frac{Z^2}{c^2} = 1. \tag{4.143}$$

The relation

$$J_1 = \frac{k^2}{a^2}; \quad J_2 = \frac{k^2}{b^2}; \quad J_3 = \frac{k^2}{c^2}, \tag{4.144}$$

expresses that the principal moments of inertia are inversely proportional to the square of the ellipsoid half-axis.

Let us have a look on a particular case.

It is assumed that all body points are located on a line (Δ), so $J_\Delta = 0$. This line is chosen as the Oz axis ($J_{zz} = 0$). The inertia products will be null and $J_{xx} = J_{yy} = J$. Thus, the inertia ellipsoid will be:

$$J_{xx}x^2 + J_{yy}y^2 = k^2, \tag{4.145}$$

or

$$J\left(x^2 + y^2\right) = k^2, \tag{4.146}$$

i.e. a rotating cylinder having the Oz-axis as the generator.

Next, we will study various practical cases where the inertia ellipsoid and its properties can be found.

(a) For a rigid body (or a material point system) at any point of the space there is an inertia ellipsoid. When the point where the matrix of moments of inertia is calculated is the center of mass of the body, it is said that the ellipsoid is the central inertia ellipsoid. The principal axes of the central inertia ellipsoid are called the inertial central axes and its principal planes are central inertia planes.

(b) Following the direction of the largest axis of the ellipsoid, the minimum moment of inertia is obtained and, after the direction of the smallest axis, we have the maximum moment of inertia. With the respect of the largest central axis of the ellipsoid we have the lowest moment of inertia that can be obtained in all space.

(c) If two principal moments of inertia are equal ($J_1 = J_2$), the ellipsoid is an ellipsoid of rotation:

$$J_1\left(X^2 + Y^2\right) + J_2 Z^2 = k^2. \tag{4.147}$$

The moment of inertia around any axis passing through the considered point, is found between the extreme values J_1 and J_2. Any axis located in the equatorial plane (XOY) is a principal axis of inertia, and the inertia moment around any of these axes is equal to J_1.

(d) If $J_1 = J_2 = J_3$, then the inertial ellipsoid becomes a sphere:

$$J_1\left(x^2 + y^2 + z^2\right) = k^2, \tag{4.148}$$

with the radius:

$$r^2 = \frac{k^2}{J_1}. \tag{4.149}$$

In this case, any axis passing through O is a principal axis of inertia, and the moment of inertia around any of these axes is equal to J_1.

(e) The moment of inertia around any axis (Δ) passing through the point considered and having the unit vector $\vec{u}(\alpha, \beta, \gamma)$ in a principal coordinate system is given by the relation:

$$J_\Delta = \alpha^2 J_1 + \beta^2 J_2 + \gamma^2 J_3. \tag{4.150}$$

If more, (Δ) is a central axis (passing through the center of mass of the rigid body), the moment around an axis (Δ_1) parallel to (Δ) and located at distance d has the expression:

$$J_{(\Delta 1)} = \alpha^2 J_1 + \beta^2 J_2 + \gamma^2 J_3 + Md^2. \tag{4.151}$$

(f) The central inertia ellipsoid of a homogeneous body follows the shape of the body. Thus, in the case of an elongated body, the central inertia ellipsoid is elongated in the same direction as the considered body;
(g) If the system has three symmetry planes perpendicular to each other, their intersections are the principal inertia axes;
(h) If a homogeneous body has an axis of symmetry, then this is a principal axis;
(i) For a rotating solid, the inertia ellipsoid is an ellipsoid of rotation around the same axis;
(j) The condition for an arbitrary ellipsoid to be an inertia ellipsoid, it must satisfy:

$$J_1 + J_2 \geq J_3; \quad J_2 + J_3 \geq J_1; \quad J_3 + J_1 \geq J_2,$$

i.e. the previously justified relationships. From here it follows:

$$\frac{1}{a^2} + \frac{1}{b^2} \geq \frac{1}{c^2}; \quad \frac{1}{b^2} + \frac{1}{c^2} \geq \frac{1}{a^2}; \quad \frac{1}{c^2} + \frac{1}{a^2} \geq \frac{1}{b^2}. \tag{4.152}$$

4.2.8 Applications

A1.

(a) If a body has two non-perpendicular symmetry planes, then any axis perpendicular to the line determined by the intersection of the two planes is a principal axis of inertia (cylindrical symmetry).
(b) If a body has three non-perpendicular symmetry planes, then any axis is a principal axis of inertia (Fig. 4.10).

Proof (a) We choose $Oz = (\pi_1) \cap (\pi_2)$ where (π_1) and (π_2) are the planes of symmetry and $Ox \in (\pi_1)$; $OX \in (\pi_2)$ so $xOz \equiv (\pi_1)$ and $XOz \equiv (\pi_2)$. The existing symmetry implies that the matrix of moments of inertia has the form:

$$[J_O] = \begin{bmatrix} J_{xx} & 0 & -J_{xz} \\ 0 & J_{yy} & 0 \\ -J_{xz} & 0 & J_{zz} \end{bmatrix}$$

Fig. 4.10 Symmetry planes

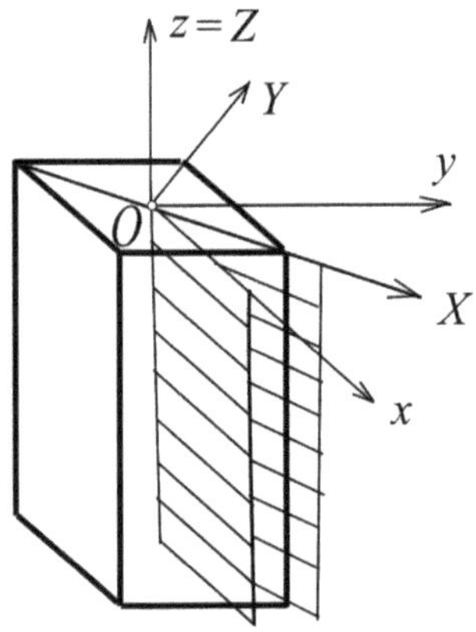

in the *Oxyz* coordinate system and the form:

$$\left[J'_O\right] = \begin{bmatrix} J'_{XX} & 0 & -J'_{Xz} \\ 0 & J'_{YY} & 0 \\ -J'_{Xz} & 0 & J'_{zz} \end{bmatrix}$$

in the *OXYz* coordinate system. The matrix that makes the transition from the *Oxyz* system to the *OXYz* reference system is:

$$[R] = \begin{bmatrix} c & s & 0 \\ -s & c & 0 \\ 0 & 0 & 1 \end{bmatrix},$$

where the abbreviations $c = \cos\theta$ and $s = \sin\theta$ are used. We therefore have:

$$\begin{aligned} \left[J'_O\right] &= [R]^T [J_O][R] \\ &= \begin{bmatrix} c^2 J_{xx} + s^2 J_{yy} & \left(J_{xx} - J_{yy}\right)sc & -cJ_{xz} \\ \left(J_{xx} - J_{yy}\right)sc & s^2 J_{xx} + c^2 J_{yy} & -sJ_{xz} \\ -cJ_{xz} & -sJ_{xz} & J_{zz} \end{bmatrix}. \end{aligned}$$

Equalizing the two expressions obtained for $\left[J'_O\right]$, by identification it results: $J_{xx} = J_{yy}$; $J_{xz} = 0$ and the matrix of moments of inertia in any system with the axis $Oz = (\pi_1) \cap (\pi_2)$ is:

$$[J_O] = \begin{bmatrix} J_1 & 0 & 0 \\ 0 & J_1 & 0 \\ 0 & 0 & J_3 \end{bmatrix}.$$

The inertia ellipsoid will be an ellipsoid of rotation and we say we have cylindrical symmetry.

(b) Either $Oz = (\pi_1) \cap (\pi_2)$ and $Oz' = (\pi_2) \cap (\pi_3)$. Compared to the $Oxyz$ system we have:

$$[J_O] = \begin{bmatrix} J_1 & 0 & 0 \\ 0 & J_2 & 0 \\ 0 & 0 & J_3 \end{bmatrix},$$

and refering to $O'x'y'z'$ we have:

$$\left[J'_O\right] = \begin{bmatrix} J'_1 & 0 & 0 \\ 0 & J'_2 & 0 \\ 0 & 0 & J'_3 \end{bmatrix}.$$

Using the result obtained in (a) in this case we will have cylindrical symmetry, so: $J_1 = J_2$ and $J'_1 = J'_2$. The invariants of the matrix of the moments of inertia are:

$$I_1 = 2J_1 + J_3 = 2J'_1 + J'_3,$$
$$I_2 = J_1^2 + 2J_1 J_3 = J_1'^2 + 2J'_3,$$
$$I_3 = J_1^2 J_3 = J_1'^2 J'_3.$$

Of the three relationships we can express J'_3 in three ways:

$$J'_3 = 2\left(J_1 - J'_1\right) + J_3 = \frac{J_1^2 + 2J_1 J_3 - J_1'^2}{2J'_1} = \frac{J_1^2 J_3}{J_1'^2}$$

with the solution: $J_1 = J'_1$; $J_3 = J'_3$. If we denote $\{e_1\} = \begin{bmatrix} \alpha & \beta & \gamma \end{bmatrix}^T$ the unit vector of the axis Ox', we have:

$$J'_1 = \{e_1\}^T [J_O]\{e_1\} = \alpha^2 J_1 + \beta^2 J_1 + \gamma^2 J_3 = J_1 + \gamma^2 (J_3 - J_1),$$

whereas $\alpha^2 + \beta^2 + \gamma^2 = 1$. But if $J'_1 = J_1$, then $J_3 = J_1$. It results:

$$[J_O] = \begin{bmatrix} J_1 & 0 & 0 \\ 0 & J_2 & 0 \\ 0 & 0 & J_3 \end{bmatrix} = J_1 \begin{bmatrix} 1 & 0 & 0 \\ 0 & 1 & 0 \\ 0 & 0 & 1 \end{bmatrix} = J_1[E],$$

and it will be the same irrespective of the chosen coordinate system. Indeed, in the arbitrarily chosen system $Ox^*y^*z^*$ we have:

$$[J_O^*] = [R^*]^T [J_O][R^*] = J_1[R^*]^T [E][R^*] = J_1[E].$$

Any axis will be the principal axis of inertia and the inertia ellipsoid will be a sphere (spherical symmetry).

A2.

(a) Determine the points of space for which the matrix of moments of inertia calculated for a rigid body has cylindrical symmetry.
(b) Determine the points of space for which the matrix of moments of inertia has spherical symmetry.

Solution

(a) We write the matrix of moments of inertia in a central coordinate system:

$$[J_C] = \begin{bmatrix} J_1 & 0 & 0 \\ 0 & J_2 & 0 \\ 0 & 0 & J_3 \end{bmatrix}.$$

Making a translation *T(x, y, z)* and applying the generalized Steiner's relationship, we obtain:

$$[J_T] = \begin{bmatrix} J_1 + M(y^2 + z^2) & -Mxy & -Mxz \\ -Mxy & J_2 + M(x^2 + z^2) & -Myz \\ -Mxz & -Myz & J_3 + M(y^2 + x^2) \end{bmatrix}.$$

To have cylindrical symmetry, we must have two equal values. The problem of eigenvalues in this case is:

$$\det \begin{bmatrix} J_1 + M(y^2 + z^2) - \lambda & -Mxy & -Mxz \\ -Mxy & J_2 + M(x^2 + z^2) - \lambda & -Myz \\ -Mzz & -Myz & J_3 + M(y^2 + x^2) - \lambda \end{bmatrix} = 0.$$

Having two equal eigenvalues, the matrix rank for which we calculate the determinant (matrix $[J_T] - \lambda[E]$) is 1. Then we will have the matrix proportional lines. Thus:

$$\begin{Bmatrix} J_1 + M(y^2 + z^2) - \lambda \\ -Mxy \\ -Mxz \end{Bmatrix} = \alpha \begin{Bmatrix} -Mxy \\ J_2 + m(x^2 + z^2) - \lambda \\ -Myz \end{Bmatrix} = \beta \begin{Bmatrix} -Mxz \\ -Myz \\ J_3 + M(y^2 + x^2) - \lambda \end{Bmatrix}.$$

Assuming $x \neq 0$, we also have $x = \alpha y$; $x = \beta z$ and the system becomes, replacing α and β in the above relations

$$\begin{aligned}
&\frac{y}{x}\left[J_1 + M\left(y^2 + z^2\right) - \lambda\right] = -Mxy, \\
&\frac{y}{x}(Mxy) = J_2 + M\left(x^2 + y^2\right) - \lambda, \\
&\frac{z}{x}\left[J_1 + M\left(y^2 + z^2\right) - \lambda\right] = -Mxz, \\
&\frac{z}{x}(Mxz) = J_3 + M\left(y^2 + x^2\right) - \lambda,
\end{aligned}$$

or

$$\begin{aligned}
&y\left[J_1 + M\left(x^2 + y^2 + z^2\right) - \lambda\right] = 0, \\
&J_2 + M\left(x^2 + y^2 + z^2\right) - \lambda = 0, \\
&z\left[J_1 + M\left(x^2 + y^2 + z^2\right) - \lambda\right] = 0, \\
&J_3 + M\left(x^2 + y^2 + z^2\right) - \lambda = 0,
\end{aligned}$$

If $y \neq 0$ or $z \neq 0$ the system becomes:

$$\begin{aligned}
&J_1 + M\left(x^2 + y^2 + z^2\right) - \lambda = 0, \\
&J_2 + M\left(x^2 + y^2 + z^2\right) - \lambda = 0, \\
&J_3 + M\left(x^2 + y^2 + z^2\right) - \lambda = 0,
\end{aligned}$$

which is incompatible, except when $J_1 = J_2 = J_3$ (spherical symmetry in C). Then the equations are checked at any point in space (so at any point we have cylindrical symmetry), and the value of two equal moments of inertia is $\lambda = J_1 + M\left(x^2 + y^2 + z^2\right)$. The matrix of moments of inertia will look like:

$$[J_T] = \begin{bmatrix} J_1 & 0 & 0 \\ 0 & J_1 + M\left(x^2 + y^2 + z^2\right) & 0 \\ 0 & 0 & J_1 + M\left(x^2 + y^2 + z^2\right) \end{bmatrix}.$$

If $y = z = 0$, the equations remain:

$$\begin{aligned}
&J_2 + M\left(x^2 + y^2 + z^2\right) - \lambda = 0, \\
&J_3 + M\left(x^2 + y^2 + z^2\right) - \lambda = 0,
\end{aligned}$$

and the system is incompatible, unless when $J_2 = J_3$. Then any point $(x, 0, 0)$ is a cylindrical symmetry point, so on the Ox-axis we have cylindrical symmetry. The equal values of the moments of inertia in regards to the other axes is $\lambda = J_1 + Mx^2$, and the matrix of moments of inertia has the form:

$$[J_T] = \begin{bmatrix} J_1 & 0 & 0 \\ 0 & J_1 + Mx^2 & 0 \\ 0 & 0 & J_1 + Mx^2 \end{bmatrix}.$$

We conclude:

- if $J_1 = J_2 = J_3$ any point is a cylindrical symmetry point;
- if $J_1 = J_2 \neq J_3$ or $J_2 = J_3 \neq J_1$ or $J_3 = J_1 \neq J_2$, then the Oz-axis and, respectively, Ox, Oy are the geometric positions of the cylindrical symmetry points.

(b) Suppose $J_1 \geq J_2 \geq J_3$. At point $T(x, y, z)$ we have:

$$\begin{aligned} [J_T] &= [J_C] + M\left[D^2\right] \\ &= \begin{bmatrix} J_1 + M\left(y^2 + z^2\right) & -Mxy & -Mxz \\ -Mxy & J_2 + M\left(z^2 + x^2\right) & -Myz \\ -Mxz & -Myz & J_3 + M\left(x^2 + y^2\right) \end{bmatrix}. \end{aligned}$$

To have spherical symmetry we must satisfy:

$$J_{xy} = J_{yz} = J_{xx} = 0,$$

and

$$J_{xx} = J_{yy} = J_{zz}.$$

Thus:

$$xy = yz = zx = 0$$

and:

$$J_1 + M\left(y^2 + z^2\right) = J_2 + M\left(z^2 + x^2\right) = J_3 + M\left(x^2 + y^2\right).$$

The first line of equality has a system of solutions:

$$\begin{aligned} x = 0,\ y = 0, \\ y = 0,\ z = 0, \\ z = 0,\ x = 0. \end{aligned}$$

Fig. 4.11 Cube

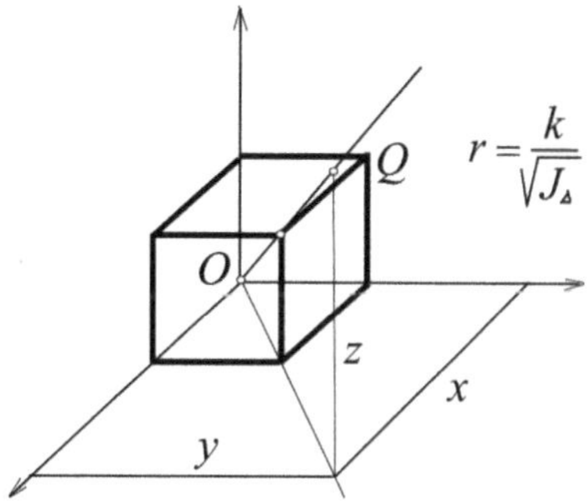

The first solution requires that: $J_1 + Mz^2 = J_2 + Mz^2 = J_3$. Thus, there is no solution unless the case when $J_1 = J_2$. Then:

$$z^2 = \frac{J_3 - J_1}{M} < 0$$

therefore unacceptable. The second solution requires that

$$J_1 = J_2 + Mx^2 = J_3 + Mx^2,$$

which only has a solution if $J_2 = J_3$. It results:

$$x = \pm\sqrt{\frac{J_1 - J_2}{M}} = \pm\sqrt{i_1^2 - i_2^2}.$$

The third solution is again unacceptable. We conclude:

- if $J_1 \neq J_2 \neq J_3$ there are no spherical symmetry points;
- if two principal central moments of inertia are equal and greater than the third, then there are two points of spherical symmetry;
- if the three principal central moments are equal, then C is the spherical symmetry point.

A3. For a cube of edge length a, represent the ellipsoid of inertia in one of its corners (see Fig. 4.11).

Solution
The cube is the rectangular paralelipiped with equal edges $a = b = c$. Thus, in the center of mass the matrix of moments of inertia will be:

$$[J_C] = \frac{Ma^2}{6}\begin{bmatrix} 1 & 0 & 0 \\ 0 & 1 & 0 \\ 0 & 0 & 1 \end{bmatrix}.$$

Writting the parallel axis theorem (Huygens-Steiner theorem) in the point O $\left(-\frac{a}{2}, -\frac{a}{2}, -\frac{a}{2}\right)$:

$$[J_O] = \frac{Ma^2}{6}\begin{bmatrix} 1 & 0 & 0 \\ 0 & 1 & 0 \\ 0 & 0 & 1 \end{bmatrix} + M\begin{bmatrix} \left(\frac{a}{2}\right)^2 + \left(\frac{a}{2}\right)^2 & -\frac{a^2}{4} & -\frac{a^2}{4} \\ -\frac{a^2}{4} & \left(\frac{a}{2}\right)^2 & -\frac{a^2}{4} \\ -\frac{a^2}{4} & -\frac{a^2}{4} & \left(\frac{a}{2}\right)^2 \end{bmatrix}$$
$$= \frac{Ma^2}{12}\begin{bmatrix} 8 & -3 & -3 \\ -3 & 8 & -3 \\ -3 & -3 & 8 \end{bmatrix}.$$

The characteristic equation reads:

$$\begin{bmatrix} 8-\lambda & -3 & -3 \\ -3 & 8-\lambda & -3 \\ -3 & -3 & 8-\lambda \end{bmatrix} = 0,$$

which gives: $\lambda_1 = 2$ and $\lambda_{2,3} = 11$. The principal moment of inertia is then:

$$J_1 = \frac{Ma^2}{12}\lambda_1 = \frac{Ma^2}{6},$$
$$J_{2,3} = \frac{Ma^2}{12}\lambda_{2,3} = \frac{11}{12}Ma^2.$$

For $\lambda = \lambda_1 = 2$, one obtains the eigenvector $\{v_1\}$ from the system:

$$\begin{bmatrix} 8-2 & -3 & -3 \\ -3 & 8-2 & -3 \\ -3 & -3 & 8-2 \end{bmatrix}\begin{Bmatrix} e_1 \\ e_2 \\ e_3 \end{Bmatrix} = \{0\},$$

with two independent equations:

$$2e_1 - e_2 - e_3 = 0,$$
$$-e_1 + 2e_2 - e_3 = 0,$$

which, together with the normalization condition $e_1^2 + e_2^2 + e_3^2 = 1$ has one of the solutions:

$$e_1 = e_2 = e_3 = \frac{\sqrt{3}}{3}.$$

Thus,

$$\{v_1\}^T = \frac{\sqrt{3}}{3}\{1\ 1\ 1\}.$$

For $\lambda = \lambda_2 = 11$ it results only one single independent equation:

$$-e_1 - e_2 - e_3 = 0.$$

To solve it, we choose $e_1 = e_2$, for example, which together with $e_1^2 + e_2^2 + e_3^2 = 1$ gives, taking $e_1 > 0$:

$$e_1 = \frac{\sqrt{6}}{6}; e_2 = \frac{\sqrt{6}}{6}; \quad e_3 = -2\frac{\sqrt{6}}{6}.$$

Thus,

$$\{v_2\}^T = \frac{\sqrt{6}}{6}\left\{1\ 1\ -2\right\}.$$

The vector $\{v_3\}$ is chosen from the condition that the coordinate system defined by the unit vectors $\{v_1\}$, $\{v_2\}$, $\{v_3\}$ to be a right hand system.

Related to the *Oxyz* system, the ellipsoid of inertia is:

$$\left[x\ y\ z\right]\begin{bmatrix} 8 & -3 & -3 \\ -3 & 8 & -3 \\ -3 & -3 & 8 \end{bmatrix}\begin{Bmatrix} x \\ y \\ z \end{Bmatrix} = k^2,$$

or

$$8\left(x^2 + y^2 + z^2\right) - 6(xy + yz + zx) = k^2,$$

and to the principal coordinate system *OXYZ*:

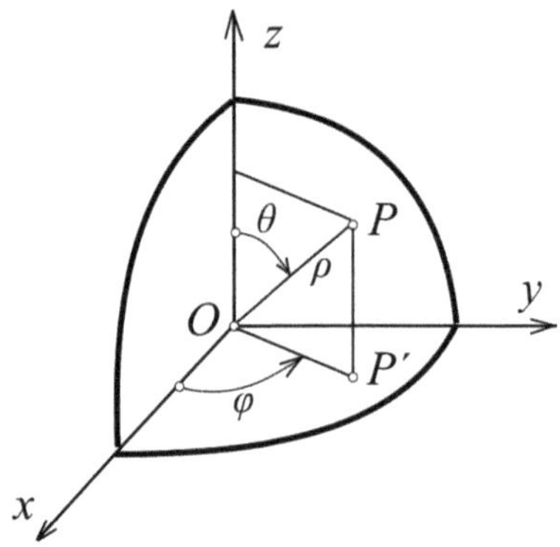

Fig. 4.12 One eighth of a sphere

$$\begin{bmatrix} X & Y & Z \end{bmatrix}\begin{bmatrix} 2 & 0 & 0 \\ 0 & 11 & 0 \\ 0 & 0 & 11 \end{bmatrix}\begin{Bmatrix} X \\ Y \\ Z \end{Bmatrix} = k^2$$

or

$$2X^2 + 11\left(Y^2 + Z^2\right) = k'^2,$$

that is, an ellipsoid of rotation.

We notice that when determining the second eigenvector, this can be arbitrarily chosen in the plane perpendicular to $\{v_1\}$ (the condition $-e_1 - e_2 - e_3 = 0$). This reflects the symmetry of the cube around the OX axis (the first principal axis of inertia).

A4. For an eighth part of a sphere with a radius r (see Fig. 4.12), determine the principal moments of inertia and represent the inertia ellipsoid related to the $Oxzy$ system. Determine the points for which the matrix of moments of inertia has spherical symmetry.

Solution

Using spherical coordinates:

$$x = \rho\sin\theta\cos\varphi,\ y = \rho\sin\theta\sin\theta,\ z = \rho\cos\theta,$$
$$\mathrm{d}V = \mathrm{d}x\mathrm{d}y\mathrm{d}z = J\mathrm{d}\rho\mathrm{d}\theta\mathrm{d}\varphi = \rho^2\sin\theta\mathrm{d}\rho\mathrm{d}\theta\mathrm{d}\varphi$$

we calculate the planar moments of inertia, which will be equal:

$$J_{xOy} = J_{yOz} = J_{zOx} = \frac{mr^2}{5}.$$

It results:

$$J_{xx} = J_{xOy} + J_{zOx} = \frac{2}{5}mr^2,$$

and analog:

$$J_{yy} = J_{zz} = \frac{2}{5}mr^2.$$

If we perform the integrations we also get:

$$J_{xy} = J_{yz} = J_{zx} = \frac{2}{5\pi}mr^2.$$

Thus,

$$[J_O] = \frac{2mr^2}{5\pi}\begin{bmatrix} \pi & -1 & -1 \\ -1 & \pi & -1 \\ -1 & -1 & \pi \end{bmatrix},$$

where m is the mass of the body. The characteristic equation

$$\begin{vmatrix} \pi - \lambda & -1 & -1 \\ -1 & \pi - \lambda & -1 \\ -1 & -1 & \pi - \lambda \end{vmatrix} = 0,$$

has the solutions $\lambda_1 = \pi - 2$; $\lambda_{2,3} = \pi + 1$. Thus, the principal moments of inertia are:

$$J_1 = \frac{2mr^2}{5\pi}(\pi - 2) \approx 0.1453\,mr^2, \quad J_2 = J_3 = \frac{2mr^2}{5\pi}(\pi + 1) \approx 0.5273mr^2.$$

The linear system

$$\begin{bmatrix} \pi - \lambda & -1 & -1 \\ -1 & \pi - \lambda & -1 \\ -1 & -1 & \pi - \lambda \end{bmatrix}\begin{Bmatrix} e_1 \\ e_2 \\ e_3 \end{Bmatrix} = 0,$$

offers for $\lambda = \lambda_1$ the following independent equations:

$$2e_1 - e_2 - e_3 = 0,$$
$$-e_1 + 2e_2 - e_3 = 0,$$

which, together with the condition: $e_1^2 + e_2^2 + e_3^2 = 1$, give:

$$e_1 = e_2 = e_3 = \pm\frac{\sqrt{3}}{3}.$$

We choose the normalized eigenvector $\{v_1\}^T = \frac{\sqrt{3}}{3}\begin{bmatrix} 1 & 1 & 1 \end{bmatrix}$.
For $\lambda = \lambda_2$ it results one single equation:

$$e_1 + e_2 + e_3 = 0,$$

to which is added the normalization condition. Choosing, for example, $e_1 = e_2$ is obtained, by taking $e_1 > 0$:

$$e_1 = \frac{\sqrt{5}}{5}; \quad e_1 = \frac{\sqrt{5}}{5}; \quad e_3 = -2\frac{\sqrt{5}}{5}.$$

Thus,

$$\{v_2\}^T = \frac{\sqrt{5}}{5}\begin{bmatrix} 1 & 1 & -2 \end{bmatrix}.$$

The third eigenvector $\{v_3\}$ is obtained by providing that the vectors $\{v_1\}$, $\{v_2\}$, $\{v_3\}$ form a right hand coordinate system. For $\lambda = \lambda_3$, proceeding as above, we will get:

$$\{v_3\}^T = \frac{\sqrt{2}}{2}\begin{bmatrix} 1 & -1 & 0 \end{bmatrix}.$$

The ellipsoid of inertia related to the $Oxyz$ system is:

$$\begin{bmatrix} x & y & z \end{bmatrix}\begin{bmatrix} \pi & -1 & -1 \\ -1 & \pi & -1 \\ -1 & -1 & \pi \end{bmatrix}\begin{Bmatrix} x \\ y \\ z \end{Bmatrix} = k^2,$$

or

$$\pi\left(x^2+y^2+z^2\right)-2(xy+yz+zx)=k^2.$$

Related to the principal coordinate system with the unit vectors $\{v_1\}, \{v_2\}, \{v_3\}$, the ellipsoid is:

$$\begin{bmatrix} X & Y & Z \end{bmatrix}\begin{bmatrix} \pi-2 & 0 & 0 \\ 0 & \pi+1 & 0 \\ 0 & 0 & \pi+1 \end{bmatrix}\begin{Bmatrix} X \\ Y \\ Z \end{Bmatrix}=k^2,$$

or

$$X^2(\pi-2)+\left(Y^2+Z^2\right)(\pi+1)=k^2,$$

i.e. an ellipsoid of rotation.

Axis *OX* is the axis of cylindrical symmetry. The matrix of moments of inertia in the center of mass $C\left(\frac{3}{8}R, \frac{3}{8}R, \frac{3}{8}R\right)$ is:

$$[J_C]=\frac{2mr^2}{5\pi}\begin{bmatrix} \pi-2 & 0 & 0 \\ 0 & \pi+1 & 0 \\ 0 & 0 & \pi+1 \end{bmatrix}-m\begin{bmatrix} 0 & 0 & 0 \\ 0 & 3\left(\frac{3}{8}\right)^2 r^2 & 0 \\ 0 & 0 & 3\left(\frac{3}{8}\right)^2 r^2 \end{bmatrix}$$

$$=\frac{2mr^2}{5\pi}\begin{bmatrix} \pi-2 & 0 & 0 \\ 0 & 1-\frac{7\pi}{128} & 0 \\ 0 & 0 & 1-\frac{7\pi}{128} \end{bmatrix}.$$

We have $J_1 > J_2 = J_3$ and that there are two points in which there is spherical symmetry, namely

$$x_{1,2}=\pm\sqrt{\frac{J_1-J_2}{3}}=\pm r\sqrt{\frac{2}{5\pi}\left(\frac{135\pi}{128}-3\right)}.$$

A5. Determine for a circular cylinder the points for which the inertial moments have spherical or cylindrical symmetry.

Solution

By doing the translation $O'(x, y, z)$ we have:

$$[J_{O'}] = \begin{bmatrix} \frac{MR^2}{4} + \frac{MH^2}{12} + M\left(y^2 + z^2\right) & -Mxy & -Mxz \\ -Mxy & \frac{MR^2}{4} + \frac{MH^2}{12} + M\left(y^2 + z^2\right) & -Myz \\ -Mxz & -Myz & \frac{MR^2}{4} + M\left(x^2 + y^2\right) \end{bmatrix}.$$

If we want to have spherical symmetry, the centrifugal moments must be zero. Thus:

$$x\,y = y\,z = z\,x = 0,$$

and the axial moments are equal. Then:

$$\frac{MR^2}{4} + \frac{MH^2}{12} + M\left(y^2 + z^2\right) = \frac{MR^2}{4} + \frac{MH^2}{12} + M\left(z^2 + x^2\right) = \frac{MR^2}{2} + Mx^2.$$

The first three equations have the solutions:

$$(a)\, x = 0,\ z = 0; \quad (b)\, y = 0,\ z = 0; \quad (c)\, x = 0,\ y = 0.$$

For case (b), from the second system, we obtain:

$$\frac{MR^2}{4} + \frac{MH^2}{12} = \frac{MR^2}{4} + \frac{MH^2}{12} + Mx^2 = \frac{MR^2}{2} + Mx^2,$$

so $x = 0$; $y = 0$; $H = R\sqrt{3}$. If the cylinder has the height $H = R\sqrt{3}$, it shows spherical symmetry and any axis passing through the center of mass is a principal axis of inertia. The principal moments of inertia are:

$$J_{xx} = J_{yy} = J_{zz} = \frac{MR^2}{2}.$$

If $x = 0$; $z = 0$, one obtains

$$\frac{MR^2}{4} + \frac{MH^2}{12} + My^2 = \frac{MR^2}{4} + \frac{MH^2}{12} = \frac{MR^2}{2} + My^2,$$

from where $y = 0$; $H = R\sqrt{3}$ (the same solution).

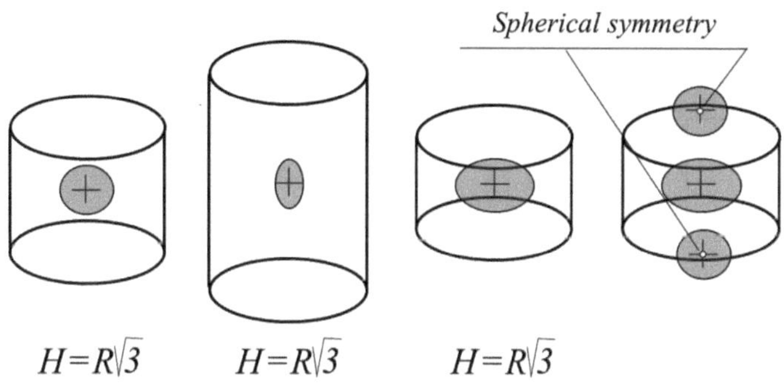

Fig. 4.13 Inertia ellipsoid for different cylinders

If $x = 0$, $y = 0$, one obtains:

$$\frac{MR^2}{4} + \frac{MH^2}{12} + Mz^2 = \frac{MR^2}{4} + \frac{MH^2}{12} + Mz^2 = \frac{MR^2}{2},$$

from where it follows:

$$z^2 = \frac{R^2}{4} - \frac{H^2}{12} = \frac{1}{12}\left(3R^2 - H^2\right).$$

If $H < R\sqrt{3}$ there are two spherical symmetry points S_1 and S_2:

$$S_1\left(0, 0, -\frac{\sqrt{3}}{6}\sqrt{3R^2 - H^2}\right); \quad S_2\left(0, 0, -\frac{\sqrt{3}}{6}\sqrt{3R^2 - H^2}\right).$$

If $H > R\sqrt{3}$ we do not have any spherical symmetry points $\left(z^2 < 0\right)$. Figure 4.13 shows ellipsoids of inertia in the cases $H = R\sqrt{3}$, $H > R\sqrt{3}$ and $H < R\sqrt{3}$.

A6. Represent the ellipsoid of inertia for the following bodies: (a) cube, sphere, regular tetrahedron, (b) cylinder, parallelepiped rectangular with square base, (c) bar.

Solution

(a) The cube, the sphere, the regular tetrahedron have spherical symmetry with respect to the center of mass, so they have equal principal moments of inertia. The matrix of the moments of inertia is:

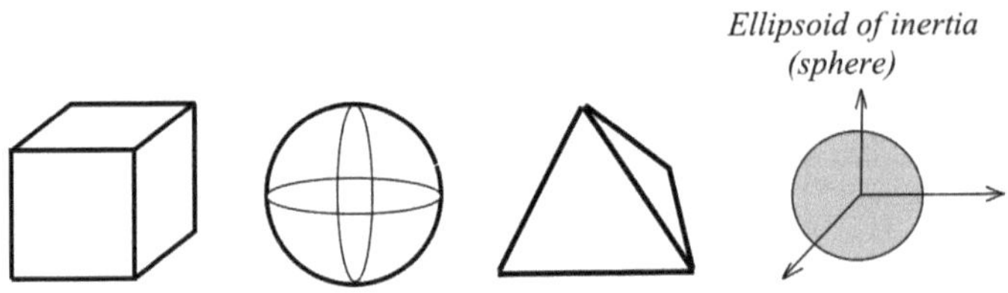

Fig. 4.14 Ellipsoids of inertia in the case of spherical symmetry

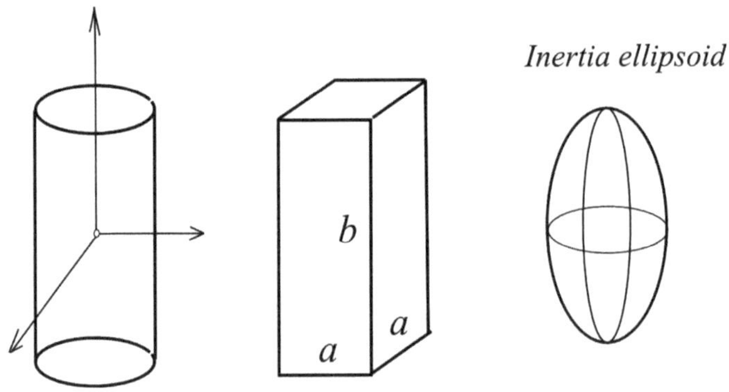

Fig. 4.15 Inertia ellipsoid when we have cylindrical symmetry

$$[J_C] = \begin{bmatrix} J & 0 & 0 \\ 0 & J & 0 \\ 0 & 0 & J \end{bmatrix},$$

so the ellipsoid of inertia will be a sphere (see Fig. 4.14).

$$J\left(x^2 + y^2 + z^2\right) = k^2.$$

(b) The cylinder and the rectangular parallelepiped with square base have cylindrical symmetry, so the matrix of moments of inertia is:

$$[J_C] = \begin{bmatrix} J_1 & 0 & 0 \\ 0 & J_1 & 0 \\ 0 & 0 & J_2 \end{bmatrix}.$$

The ellipsoid of inertia will be an ellipsoid of rotation (see Fig. 4.15).

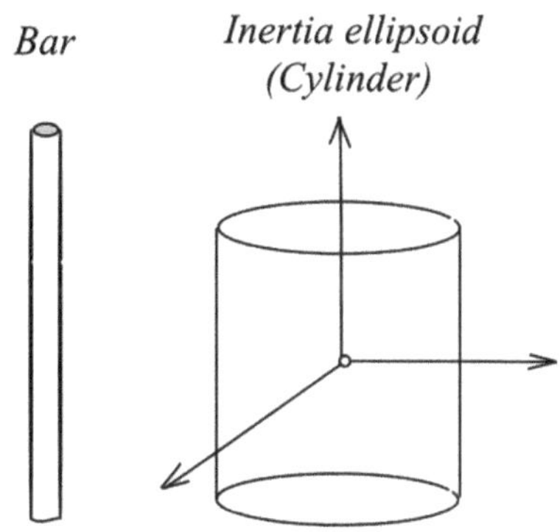

Fig. 4.16 Inertia ellipsoid for a bar

(c) By placing the Oz axis along the bar, we have:

$$[J_C] = \begin{bmatrix} J & 0 & 0 \\ 0 & J & 0 \\ 0 & 0 & 0 \end{bmatrix}.$$

Thus, the ellipsoid of inertia degenerates into a cylinder (see Fig. 4.16)

$$J\left(x^2 + y^2\right) = k^2.$$

A7. Let us find the points of the spherical symmetry for a rectangular cone.

Solution

In the center of mass C of the cone $C\left(0, 0, \frac{H}{4}\right)$ the inertia moment matrix is (see Vlase 2004):

$$\begin{aligned}[J_C] &= [J_O] - M\left[D^2\right] \\ &= M\begin{bmatrix} \frac{3R^2}{20} + \frac{H^2}{10} & 0 & 0 \\ 0 & \frac{3R^2}{20} + \frac{H^2}{10} & 0 \\ 0 & 0 & \frac{3R^2}{10} \end{bmatrix} - M\begin{bmatrix} \frac{H^2}{16} & 0 & 0 \\ 0 & \frac{H^2}{16} & 0 \\ 0 & 0 & 0 \end{bmatrix} \\ &= \frac{3MR^2}{20}\begin{bmatrix} 1 + \frac{H^2}{4R^2} & 0 & 0 \\ 0 & 1 + \frac{H^2}{4R^2} & 0 \\ 0 & 0 & 2 \end{bmatrix}.\end{aligned}$$

The Oz-axis is a cylindrical symmetry axis. We are in the condition of the application A2 when $J_1 = J_2$. It results the spherical symmetry points:

$$z_{1,2} = \frac{H}{4} \pm \sqrt{\frac{J_3 - J_1}{M}} = \frac{H}{4} \pm \sqrt{\frac{3R^2}{20}\left(1 - \frac{H^2}{4R^2}\right)}.$$

Thus:

- if $1 - \frac{H^2}{4R^2} > 0$, that is $H < 2R$, we have two points of spherical symmetry;
- if $1 - \frac{H^2}{4R^2} = 0$, that is $H = 2R$, the center of mass is a spherical symmetry point;
- if $1 - \frac{H^2}{4R^2} < 0$, i.e. $H > 2R$, we do not have spherical symmetry points.

4.2.9 Geometrical Moments of Inertia

For a rigid body, the moment of inertia about an axis (Δ) is calculated with the relation:

$$J_\Delta = \int \delta^2 \, \mathrm{d}m.$$

If ρ is the mass density of the homogeneous body and V is its volume, we have:

$$J_\Delta = \int \delta^2 \rho \mathrm{d}V = \rho I_\Delta, \tag{4.153}$$

where the quantity $I_\Delta = \int \delta^2 \mathrm{d}V$ is called geometrical moment of axial inertia. Generally, for homogeneous bodies for each moment of inertia there is a correspondent quantity, the geometric moment of inertia. The both quantities are linked by the relationship:

$$J = \rho I. \tag{4.154}$$

If the body is a homogeneous plane plate having the area A, the moment of geometric axial inertia is:

$$I_\Delta = \int \delta^2 \, \mathrm{d}A. \tag{4.155}$$

If the body is a homogeneous line then:

$$I_\Delta = \int \delta^2 \, \mathrm{d}s. \tag{4.156}$$

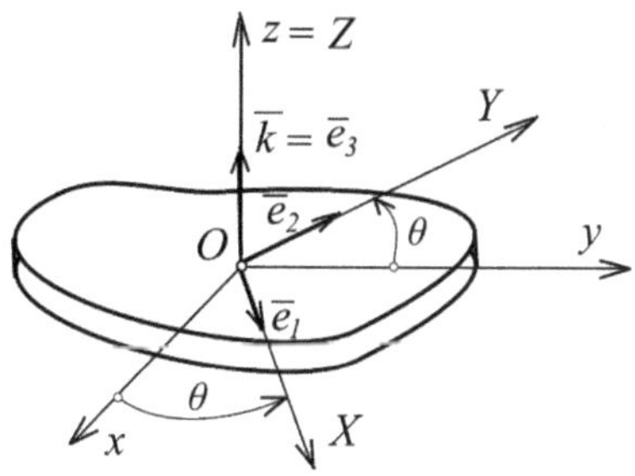

Fig. 4.17 Plate

The inertia radius for a homogeneous body is given by:

$$i_{\Delta} = \sqrt{\frac{J_{\Delta}}{M}} = \sqrt{\frac{I_{\Delta}}{V}}. \tag{4.157}$$

In an orthogonal coordinate system we have:

$$i_x = \sqrt{\frac{J_{xx}}{M}} = \sqrt{\frac{I_{xx}}{V}}, \quad i_y = \sqrt{\frac{J_{yy}}{M}} = \sqrt{\frac{I_{yy}}{V}}, \quad i_z = \sqrt{\frac{J_{zz}}{M}} = \sqrt{\frac{I_{zz}}{V}}. \tag{4.158}$$

If we consider a flat plate of area A, then the inertia radius is given by:

$$i_x = \sqrt{\frac{I_{xx}}{A}}; \quad i_y = \sqrt{\frac{I_{yy}}{A}}. \tag{4.159}$$

4.2.10 Moment of Inertia of Planar Plates

If the body mass is considered to be distributed in a plane ($z = 0$, see Fig. 4.17), we have:

$$J_{xx} + J_{yy} = \int \left(y^2 + z^2\right)\mathrm{d}m + \int \left(x^2 + z^2\right)\mathrm{d}m = \int \left(x^2 + y^2\right)\mathrm{d}m = J_{zz}. \tag{4.160}$$

Thus, we have in the case of flat plates the relation: $J_{zz} = J_{xx} + J_{yy}$. Further, we have:

$$J_{xz} = \int xz\,\mathrm{d}m = 0 \quad \text{and} \quad J_{yz} = \int yz\,\mathrm{d}m = 0. \tag{4.161}$$

The matrix of the moments of inertia has the form:

$$[J_O] = \begin{bmatrix} J_{xx} & -J_{xy} & 0 \\ -J_{xy} & J_{yy} & 0 \\ 0 & 0 & J_{zz} \end{bmatrix}. \tag{4.162}$$

If we consider an *OXYZ* coordinate system, rotated with the angle θ around the axis Oz, the rotation matrix is:

$$[R] = \begin{bmatrix} \{e_1\} \; \{e_2\} \; \{e_3\} \end{bmatrix} = \begin{bmatrix} c & -s & 0 \\ s & c & 0 \\ 0 & 0 & 1 \end{bmatrix}. \tag{4.163}$$

The matrix of moments of inertia becomes, in the new reference system:

$$[J_O]' = \begin{bmatrix} c & s & 0 \\ -s & c & 0 \\ 0 & 0 & 1 \end{bmatrix} \begin{bmatrix} J_{xx} & -J_{xy} & 0 \\ -J_{xy} & J_{yy} & 0 \\ 0 & 0 & 1 \end{bmatrix} \begin{bmatrix} c & -s & 0 \\ s & c & 0 \\ 0 & 0 & 1 \end{bmatrix}$$
$$= \begin{bmatrix} J_{xx}c^2 - 2J_{xy}sc + J_{yy}s^2 & sc\left(J_{yy} - J_{xx}\right) + J_{xy}\left(s^2 - c^2\right) & 0 \\ sc\left(J_{yy} - J_{xx}\right) + J_{xy}\left(s^2 - c^2\right) & J_{xx}s^2 - 2J_{xy}sc + J_{yy}c^2 & 0 \\ 0 & 0 & J_{zz} \end{bmatrix}. \tag{4.164}$$

Thus, for example:

$$J_{XX} = J_{xx}c^2 - 2J_{xy}sc + J_{yy}s^2. \tag{4.165}$$

In a principal axes system the ellipsoid of inertia has the classical form (4.141). Since the principal semi-axes of the ellipsoid of inertia are given by Eq. (4.142) the condition (4.160) leads to:

$$\frac{1}{a^2} + \frac{1}{b^2} = \frac{1}{c^2}. \tag{4.166}$$

The intersection of the Poinsot's ellipsoid with the *Oxy* plane is an ellipse, called the ellipse of inertia of the plate:

$$\begin{bmatrix} x & y \end{bmatrix} = \begin{bmatrix} J_{xx} & -J_{xy} \\ -J_{xy} & J_{yy} \end{bmatrix} \begin{Bmatrix} x \\ y \end{Bmatrix} = k^2, \tag{4.167}$$

or

$$J_{xx}x^2 + J_{yy}y^2 - 2J_{xy}xy = k^2. \tag{4.168}$$

In a principal axes system the ellipse of inertia has the form:

$$J_1 X^2 + J_2 Y^2 = k^2, \tag{4.169}$$

or, dividing it with the invariant $I_2 = J_1 J_2$:

$$\frac{X^2}{J_2} + \frac{Y^2}{J_1} = \frac{k^2}{J_1 J_2}. \tag{4.170}$$

Taking into account the relationships between the moments of inertia and the radius of inertia, i.e. $J_1 = m i_1^2$ and $J_2 = m i_2^2$, it results:

$$\frac{X^2}{i_2^2} + \frac{Y^2}{i_1^1} = \frac{k^2}{m} = k'^2. \tag{4.171}$$

If we choose the constant k' equal to 1, it is possible to represent, on a certain scale, the ellipse of inertia:

$$\left(\frac{X}{i_2}\right)^2 + \left(\frac{Y}{i_1}\right)^2 = 1. \tag{4.172}$$

Reference

Vlase, S.: Mechanics. Dynamics. Transylvania University of Brasov Press, Brasov (2004)

Chapter 5
Strain and Stress

5.1 Strain Tensor

5.1.1 Deformations

Considering a continous solid, a deformation of this means that a curve changes its length when it changes its position.

If the length of any curve of the solid does not change we say that we have a rigid motion of the solid. For an analysis of the deformation of a body the undeformed configuration is considered as the reference configuration and the current configuration is the deformed configuration. The components of the position vector of a point (particle) in the reference configuration system are called the material or reference coordinates (see Lurie and Belyaev 2005; Maceri 2010; Sharpe 2008; Landau and Lifchitz 1967; Teodorescu 1970, 2006).

Under the action of a system of forces in equilibrium, a solid body is deformed so that it changes the shape and volume. The point $P(x, y, z)$ having the position vector $\{r\}$ passes at point $P'(x', z', y')$ with the position vector $\{r'\}$. Similarly, $Q(x + dx, y + dy, z + dz)$ infinitely adjacent to P passes to $Q'(x' + dx', y' + dy', z' + dz')$. If we note with $\{u\}$ the deformation vector of P in P', we can write the relation:

$$\{r'\} = \{r\} + \{u\}, \tag{5.1}$$

or, by components:

$$\begin{Bmatrix} x' \\ y' \\ z' \end{Bmatrix} = \begin{Bmatrix} x \\ y \\ z \end{Bmatrix} + \begin{Bmatrix} u_x(x, y, z) \\ u_y(x, y, z) \\ u_z(x, y, z) \end{Bmatrix}. \tag{5.1'}$$

S. Vlase et al., *Eigenvalue and Eigenvector Problems in Applied Mechanics*, Advanced Structured Materials 96, https://doi.org/10.1007/978-3-030-00991-5_5

The spatial (i.e. three-dimensional) deformation state is the general case of deformation of any elastic body (Gould 2013).

If the domain D is occupied by the solid body before deformation (at time $t = t_o$) it becomes, after deformation, the domain D' (at time t). It is considered that the transition from the initial position to the final position is made continuously. The new coordinates are functions of the old coordinates and time:

$$x' = x'(x, y, z, t), \quad y' = y'(x, y, z, t), \quad z' = z'(x, y, z, t). \tag{5.2}$$

The reverse transformation, i.e.

$$x = x(x', y', z', t), \quad y = y(x', y', z', t), \quad z = z(x', y', z', t). \tag{5.3}$$

is done uniquely. It is said in this case that we are dealing with uniform displacements (there is also the case of multiple linked domains where we can have multiforme displacements).

In order that functions (5.3) are uniquely determined in a vecinity of the considered point, functions (5.2) must be of class C^1 in the domain D and have a non-zero Jacobian determinant:

$$\frac{\partial(x', y', z')}{\partial(x, y, z)} = \begin{vmatrix} \frac{\partial x'}{\partial x} & \frac{\partial x'}{\partial y} & \frac{\partial x'}{\partial z} \\ \frac{\partial y'}{\partial x} & \frac{\partial y'}{\partial y} & \frac{\partial y'}{\partial z} \\ \frac{\partial z'}{\partial x} & \frac{\partial z'}{\partial y} & \frac{\partial z'}{\partial z} \end{vmatrix} \neq 0, \tag{5.4}$$

in the domain D. This hypothesis is called the continuity axiom and expresses the indestructibility of matter. A domain corresponding to a finite positive volume cannot be converted to a zero or infinite volume. This property expresses the impenetrability of matter.

5.1.2 *Lagrangian and Eulerian Description*

Two methods are usually used to study the deformation of a solid body (continuum medium). The first description is made in terms of the reference (material) coordinates and is called the Lagrangian description or material description. The second is made in terms of the spatial coordinates and is called the Eulerian description or spatial description.

The following two hypotheses will be used below:

- the deformations of an elastic body are continuous, i.e. there are no dislocations (a closed curve at any instant will form a closed curve at any subsequent time and a closed surface will always form, at any subsequent time, a closed surface);

- deformations of the elastic body are very small.

Lagrange's method. We will study the transformation of an infinitesimal vector $\mathrm{d}\bar{r} = \overline{PQ}$ during deformation. We have:

$$\{\mathrm{d}r'\} = \{\mathrm{d}r\} + \{\mathrm{d}u\}. \tag{5.5}$$

Further, if one takes into account the definition of differentiation, it can be written:

$$\{\mathrm{d}u\} = \begin{Bmatrix} \mathrm{d}u_x \\ \mathrm{d}u_y \\ \mathrm{d}u_z \end{Bmatrix} = \begin{bmatrix} \frac{\partial u_x}{\partial x} & \frac{\partial u_x}{\partial y} & \frac{\partial u_x}{\partial z} \\ \frac{\partial u_y}{\partial x} & \frac{\partial u_y}{\partial y} & \frac{\partial u_y}{\partial z} \\ \frac{\partial u_z}{\partial x} & \frac{\partial u_z}{\partial y} & \frac{\partial u_z}{\partial z} \end{bmatrix} \begin{Bmatrix} \mathrm{d}x \\ \mathrm{d}y \\ \mathrm{d}z \end{Bmatrix} = [\partial]\{\mathrm{d}r\}, \tag{5.6}$$

where we used the abbreviation:

$$[\partial] = \begin{bmatrix} \frac{\partial u_x}{\partial x} & \frac{\partial u_x}{\partial y} & \frac{\partial u_x}{\partial z} \\ \frac{\partial u_y}{\partial x} & \frac{\partial u_y}{\partial y} & \frac{\partial u_y}{\partial z} \\ \frac{\partial u_z}{\partial x} & \frac{\partial u_z}{\partial y} & \frac{\partial u_z}{\partial z} \end{bmatrix}. \tag{5.7}$$

In this case it can be written:

$$\{\mathrm{d}r'\} = \{\mathrm{d}r\} + [\partial]\{\mathrm{d}r\} = ([E] + [\partial])\{\mathrm{d}r\}. \tag{5.8}$$

We take the distance between neighboring points P and Q ($\mathrm{d}\bar{r} = \overline{PQ}$, $\mathrm{d}\bar{r}' = \overline{P'Q'}$) as a measure of the deformation. It is obtained:

$$\begin{aligned}(\mathrm{d}r')^2 &= \{\mathrm{d}r'\}^{\mathrm{T}}\{\mathrm{d}r'\} = \{\mathrm{d}r'\}^{\mathrm{T}}([E] + [\partial]^T)([E] + [\partial])\{\mathrm{d}r'\} \\ &= \{\mathrm{d}r\}^{\mathrm{T}}([E] + [\partial] + [\partial]^T + [\partial]^T[\partial])\{\mathrm{d}r'\}.\end{aligned} \tag{5.9}$$

It is noted: $2[T_L] = [\partial] + [\partial]^T + [\partial]^T[\partial]$ where $[T_{\mathrm{L}}]$ is the Lagrange deformation tensor:

$$[T_{\mathrm{L}}] = \begin{bmatrix} e_{xx} & \frac{1}{2}e_{xy} & \frac{1}{2}e_{xz} \\ \frac{1}{2}e_{xy} & e_{yy} & \frac{1}{2}e_{yz} \\ \frac{1}{2}e_{xz} & \frac{1}{2}e_{yz} & e_{zz} \end{bmatrix}, \tag{5.10}$$

where the components of this tensor are:

$$
\begin{aligned}
e_{xx} &= \frac{\partial u_x}{\partial x} + \frac{1}{2}\left[\left(\frac{\partial u_x}{\partial x}\right)^2 + \left(\frac{\partial u_y}{\partial x}\right)^2 + \left(\frac{\partial u_z}{\partial x}\right)^2\right],\\
e_{yy} &= \frac{\partial u_y}{\partial y} + \frac{1}{2}\left[\left(\frac{\partial u_x}{\partial y}\right)^2 + \left(\frac{\partial u_y}{\partial y}\right)^2 + \left(\frac{\partial u_z}{\partial y}\right)^2\right],\\
e_{zz} &= \frac{\partial u_z}{\partial z} + \frac{1}{2}\left[\left(\frac{\partial u_x}{\partial z}\right)^2 + \left(\frac{\partial u_y}{\partial z}\right)^2 + \left(\frac{\partial u_z}{\partial z}\right)^2\right],\\
e_{xy} &= e_{yx} = \frac{\partial u_x}{\partial y} + \frac{\partial u_y}{\partial x} + \frac{\partial u_x}{\partial x}\frac{\partial u_x}{\partial y} + \frac{\partial u_y}{\partial x}\frac{\partial u_y}{\partial y} + \frac{\partial u_z}{\partial x}\frac{\partial u_z}{\partial y},\\
e_{yz} &= e_{zy} = \frac{\partial u_y}{\partial z} + \frac{\partial u_z}{\partial y} + \frac{\partial u_y}{\partial y}\frac{\partial u_y}{\partial z} + \frac{\partial u_z}{\partial y}\frac{\partial u_z}{\partial z} + \frac{\partial u_x}{\partial y}\frac{\partial u_x}{\partial z},\\
e_{zx} &= e_{xz} = \frac{\partial u_z}{\partial x} + \frac{\partial u_x}{\partial z} + \frac{\partial u_z}{\partial z}\frac{\partial u_z}{\partial x} + \frac{\partial u_x}{\partial z}\frac{\partial u_x}{\partial x} + \frac{\partial u_y}{\partial z}\frac{\partial u_y}{\partial x}.
\end{aligned}
\tag{5.11}
$$

We have:

$$
\begin{aligned}
\mathrm{d}r^2 &= \mathrm{d}s^2 = \mathrm{d}x^2 + \mathrm{d}y^2 + \mathrm{d}z^2\\
\mathrm{d}r'^2 &= \mathrm{d}s'^2 = dx'^2 + \mathrm{d}y'^2 + \mathrm{d}z'^2 = (\mathrm{d}x + \mathrm{d}u)^2 + (\mathrm{d}y + \mathrm{d}v)^2 + (\mathrm{d}z + \mathrm{d}w)^2\\
&= (1 + 2e_{xx})\mathrm{d}x + \left(1 + 2e_{yy}\right)\mathrm{d}y + (1 + 2e_{zz})\mathrm{d}z\\
&\quad + 2e_{yz}dydz + 2e_{zx}dzdx + 2e_{xy}dxdy.
\end{aligned}
\tag{5.12}
$$

The variation of the distance between points P and Q is given by:

$$
\frac{1}{2}\left(\mathrm{d}s'^2 - \mathrm{d}s^2\right) = \mathrm{e_{xx}}\mathrm{d}x^2 + \mathrm{e_{yy}}\mathrm{d}y^2 + \mathrm{e_{zz}}\mathrm{d}z^2 + 2\mathrm{e_{yz}}\mathrm{d}y\mathrm{d}z + 2\mathrm{e_{zx}}\mathrm{d}z\mathrm{d}x + 2\mathrm{e_{xy}}\mathrm{d}x\mathrm{d}y.
\tag{5.13}
$$

Euler's method. We will study the transformation of an infinitesimal vector $\mathrm{d}\bar{r} = \overline{PQ}$ during deformation. We have:

$$
\left\{\mathrm{d}r'\right\} = \{\mathrm{d}r\} + \{\mathrm{d}u\}.
\tag{5.14}
$$

Further, if one takes into account the definition of differentiation, one can write:

$$
\{\mathrm{d}u\} = \begin{Bmatrix} \mathrm{d}u_x \\ \mathrm{d}u_y \\ \mathrm{d}u_z \end{Bmatrix} = \begin{bmatrix} \frac{\partial u_x}{\partial x'} & \frac{\partial u_x}{\partial y'} & \frac{\partial u_x}{\partial z'} \\ \frac{\partial u_y}{\partial x'} & \frac{\partial u_y}{\partial y'} & \frac{\partial u_y}{\partial z'} \\ \frac{\partial u_z}{\partial x'} & \frac{\partial u_z}{\partial y'} & \frac{\partial u_z}{\partial z'} \end{bmatrix} \begin{Bmatrix} \mathrm{d}x' \\ \mathrm{d}y' \\ \mathrm{d}z' \end{Bmatrix} = [\partial']\{\mathrm{d}r'\},
\tag{5.15}
$$

where we used the abbreviation

$$[\partial'] = \begin{bmatrix} \frac{\partial u_x}{\partial x'} & \frac{\partial u_x}{\partial y'} & \frac{\partial u_x}{\partial z'} \\ \frac{\partial u_y}{\partial x'} & \frac{\partial u_y}{\partial y'} & \frac{\partial u_y}{\partial z'} \\ \frac{\partial u_z}{\partial x'} & \frac{\partial u_z}{\partial y'} & \frac{\partial u_z}{\partial z'} \end{bmatrix}. \tag{5.16}$$

In this case it can be written:

$$\{\mathrm{d}r\} = \left\{\mathrm{d}r'\right\} - [\partial]\left\{\mathrm{d}r'\right\} = ([E] + [\partial])\left\{\mathrm{d}r'\right\}, \tag{5.17}$$

where $[E]$ is the unit matrix.

We take the distance between the neighboring points P and Q ($\mathrm{d}\bar{r} = \overline{PQ},\ \mathrm{d}\bar{r}' = \overline{P'Q'}$) as a measure of the deformation. It is obtained:

$$\begin{aligned}(\mathrm{d}r)^2 &= \{\mathrm{d}r\}^{\mathrm{T}}\{\mathrm{d}r\} = \{\mathrm{d}r\}^{\mathrm{T}}([E] - [\partial]^T)([E] - [\partial])\{\mathrm{d}r\} \\ &= \left\{\mathrm{d}r'\right\}^{\mathrm{T}}([E] - [\partial] - [\partial]^T + [\partial]^T[\partial])\left\{\mathrm{d}r'\right\}. \end{aligned} \tag{5.18}$$

It is denoted $2[T_{\mathrm{E}}] = [\partial] + [\partial]^T - [\partial]^T[\partial]$, where $[T_{\mathrm{E}}]$ is called the Eulerian deformation tensor

$$[T_{\mathrm{E}}] = \begin{bmatrix} e'_{xx} & \frac{1}{2}e'_{xy} & \frac{1}{2}e'_{xz} \\ \frac{1}{2}e'_{xy} & e'_{yy} & \frac{1}{2}e'_{yz} \\ \frac{1}{2}e'_{xz} & \frac{1}{2}e'_{yz} & e'_{zz} \end{bmatrix}, \tag{5.19}$$

where the components of this tensor are:

$$\begin{aligned}
e_{xx} &= \frac{\partial u_x}{\partial x'} - \frac{1}{2}\left[\left(\frac{\partial u_x}{\partial x'}\right)^2 + \left(\frac{\partial u_y}{\partial x'}\right)^2 + \left(\frac{\partial u_z}{\partial x'}\right)^2\right], \\
e_{yy} &= \frac{\partial u_y}{\partial y'} - \frac{1}{2}\left[\left(\frac{\partial u_x}{\partial y'}\right)^2 + \left(\frac{\partial u_y}{\partial y'}\right)^2 + \left(\frac{\partial u_z}{\partial y'}\right)^2\right], \\
e_{zz} &= \frac{\partial u_z}{\partial z'} - \frac{1}{2}\left[\left(\frac{\partial u_x}{\partial z'}\right)^2 + \left(\frac{\partial u_y}{\partial z'}\right)^2 + \left(\frac{\partial u_z}{\partial z'}\right)^2\right], \\
e_{xy} = e_{yx} &= \frac{\partial u_x}{\partial y'} + \frac{\partial u_y}{\partial x'} - \left(\frac{\partial u_x}{\partial x'}\frac{\partial u_x}{\partial y'} + \frac{\partial u_y}{\partial x'}\frac{\partial u_y}{\partial y'} + \frac{\partial u_z}{\partial x'}\frac{\partial u_z}{\partial y'}\right), \\
e_{yz} = e_{zy} &= \frac{\partial u_y}{\partial z'} + \frac{\partial u_z}{\partial y'} - \left(\frac{\partial u_y}{\partial y'}\frac{\partial u_y}{\partial z'} + \frac{\partial u_z}{\partial y'}\frac{\partial u_z}{\partial z'} + \frac{\partial u_x}{\partial y'}\frac{\partial u_x}{\partial z'}\right), \\
e_{zx} = e_{xz} &= \frac{\partial u_z}{\partial x'} + \frac{\partial u_x}{\partial z'} - \left(\frac{\partial u_z}{\partial z'}\frac{\partial u_z}{\partial x'} + \frac{\partial u_x}{\partial z'}\frac{\partial u_x}{\partial x'} + \frac{\partial u_y}{\partial z'}\frac{\partial u_y}{\partial x'}\right).
\end{aligned} \tag{5.20}$$

We have:

$$\begin{aligned}
dr'^2 &= ds'^2 = dx'^2 + dy'^2 + dz'^2, \\
dr^2 &= ds^2 = dx^2 + dy^2 + dz^2 = \left(dx' - du\right)^2 + \left(dy' - dv\right)^2 + \left(dz' - dw\right)^2 \\
&= (1 - 2e_{xx})dx + \left(1 - 2e_{yy}\right)dy + (1 - 2e_{zz})dz \\
&\quad - 2e_{yz}dydz - 2e_{zx}dzdx - 2e_{xy}dxdy. \qquad (5.21)
\end{aligned}$$

The variation of the distance between points P and Q is given by:

$$\frac{1}{2}\left(ds'^2 - ds^2\right) = e'_{xx}dx'^2 + e'_{yy}dy'^2 + e'_{zz}dz'^2 + 2e'_{yz}dy'dz' + 2e'_{zx}dz'dx' + 2e'_{xy}dx'dy'. \qquad (5.22)$$

5.1.3 Strain

Consider the quantity:

$$\varepsilon'_n = \frac{ds' - ds}{ds}, \qquad (5.23)$$

which is called the strain (specific length) of a linear element in direction of the vector $\vec{n}$. If $\varepsilon_n > 0$ we have an elongation (tensile strain) and if $\varepsilon_n < 0$ we have a compression (compressive strain).

Relation (5.13) gives us in Lagrangian coordinates:

$$\begin{aligned}
\varepsilon_n\left(1 + \frac{1}{2}\varepsilon_n\right) &= e_{xx}\cos^2(n, x) + e_{yy}\cos^2(n, y) + e_{zz}\cos^2(n, z) \\
&\quad + e_{yz}\cos(n, y)\cos(n, z) + e_{zx}\cos(n, z)\cos(n, x) \\
&\quad + e_{xy}\cos(n, x)\cos(n, y), \qquad (5.24)
\end{aligned}$$

where the following directional cosines are used:

$$\cos(n, x) = \frac{dx}{ds}; \quad \cos(n, y) = \frac{dy}{ds}; \quad \cos(n, z) = \frac{dz}{ds}. \qquad (5.25)$$

If we make $\vec{n}$ to be the Ox-axis, we get:

$$\varepsilon_{xx}\left(1 + \frac{1}{2}\varepsilon_{xx}\right) = e_x, \qquad (5.26)$$

from where the admissible solution is obtained:

$$\varepsilon_{xx} = \sqrt{1 + 2e_{xx}} - 1. \qquad (5.27)$$

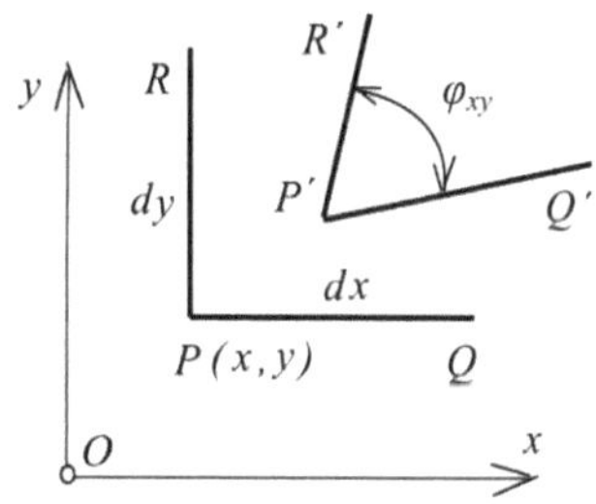

Fig. 5.1 Change of the right angle in Lagrangian coordinates

Hence, the significance of the magnitudes ε_{xx}, ε_{yy}, ε_{zz}, which are the strains along the directions of the coordinate axes, are expressed according to the tensor $[T_L]$.

Let us now consider two segments $\overrightarrow{PQ}$ and $\overrightarrow{PR}$ parallel to the coordinate axes Oy and Oz before deformation (thus, perpendicular to each other). After deformation they will form an angle, i.e.

$$\gamma_{yz} = \frac{\pi}{2} - \varphi_{yz}, \tag{5.28}$$

which defines the variation of the right angle formed before deformation by these two segments. Thus, γ_{xy}, γ_{yz}, and γ_{zx} it also characterize specific angular deformations and are called the shear strains. They are positive when they represent a decrease in the right angle and correspond to a positive tangential shear stress and negative if it is an increase in the right angle and correspond to a negative tangential shear stress.

Lagrangian coordinates it can be written (see Fig. 5.1):

$$\sin\,\gamma_{yz} = \frac{e_{yz}}{\sqrt{1+2e_{yy}}\sqrt{1+2e_{zz}}}, \tag{5.29}$$

and, similarly:

$$\sin\,\gamma_{zx} = \frac{e_{zx}}{\sqrt{1+2e_{zz}}\sqrt{1+2e_{xx}}}, \tag{5.30}$$

$$\sin\,\gamma_{xy} = \frac{e_{xy}}{\sqrt{1+2e_{zz}}\sqrt{1+2e_{xx}}}. \tag{5.31}$$

The change ratio of a very small volume in the strained and in the unstrained state can be written in the following form:

$$\varepsilon_v = (1+\varepsilon_{xx})\left(1+\varepsilon_{yy}\right)(1+\varepsilon_{zz}) - 1 = \sqrt{1+2e_{xx}}\sqrt{1+2e_{yy}}\sqrt{1+2e_{zz}} - 1, \tag{5.32}$$

Or

$$\varepsilon_v + 1 = \frac{\partial(x', y', z')}{\partial(x, y, z)} = \begin{vmatrix} 1 + \frac{\partial u}{\partial x} & \frac{\partial u}{\partial y} & \frac{\partial u}{\partial z} \\ \frac{\partial v}{\partial x} & 1 + \frac{\partial v}{\partial y} & \frac{\partial v}{\partial z} \\ \frac{\partial w}{\partial x} & \frac{\partial w}{\partial y} & \frac{\partial w}{\partial z} \end{vmatrix}. \tag{5.33}$$

In Eulerian coordinates, a similar strain (specific length) can also be defined:

$$\varepsilon'_n = \frac{\mathrm{d}s' - \mathrm{d}s}{\mathrm{d}s'}. \tag{5.34}$$

If $\varepsilon'_n > 0$ we have a tensile strain and if $\varepsilon'_n < 0$ we have a compressive strain. Relation (5.34) offers:

$$\begin{aligned} \varepsilon'_n\left(-1 + \frac{1}{2}\varepsilon'_n\right) &= e'_{xx}\cos^2(n, x') + e'_{yy}\cos^2(n, y') + e'_{zz}\cos^2(n, z') \\ &+ e'_{yz}\cos(n, y')\cos(n, z') + e'_{zx}\cos(n, z')\cos(n, x') \\ &+ e'_{xy}\cos(n, x')\cos(n, y'), \end{aligned} \tag{5.35}$$

where we used for the directional cosines the following expression:

$$\cos(n, x') = \frac{dx'}{ds}; \quad \cos(n, y') = \frac{dy'}{ds}; \quad \cos(n, z') = \frac{dz'}{ds}. \tag{5.36}$$

If the normal vector $\vec{n}$ is the Ox-axis, one obtains:

$$\varepsilon'_{xx}\left(-1 + \frac{1}{2}\varepsilon'_{xx}\right) = e'_{x}, \tag{5.37}$$

from where:

$$\varepsilon'_{xx} = 1 - \sqrt{1 - 2e'_{xx}}. \tag{5.38}$$

The strains $\varepsilon'_{xx}, \varepsilon'_{yy}, \varepsilon'_{zz}$, expressed in terms of the Eulerian strain tensor $[T_{\mathrm{E}}]$ result as:

The shear strains are given by the following relations (see Fig. 5.2):

$$\sin\gamma'_{xy} = \frac{e'_{xy}}{\sqrt{1 + 2e'_{xx}}\sqrt{1 + 2e'_{yy}}}, \tag{5.39}$$

$$\sin\gamma'_{yz} = \frac{e'_{yz}}{\sqrt{1 + 2e'_{yy}}\sqrt{1 + 2e'_{zz}}}, \tag{5.40}$$

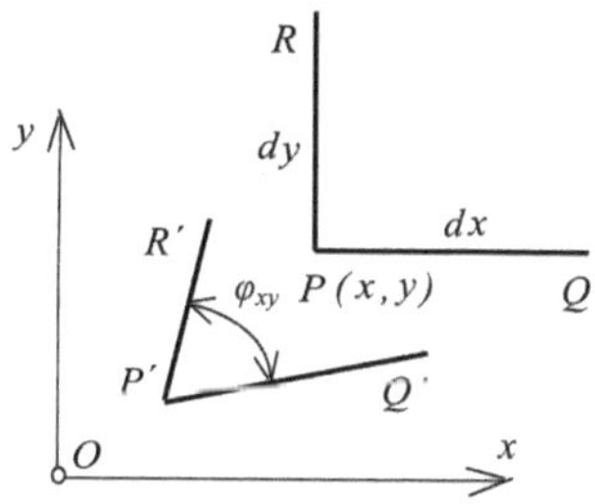

Fig. 5.2 Change of the right angle in Eulerian coordinates

$$\sin \gamma'_{zx} = \frac{e'_{zx}}{\sqrt{1 + 2e'_{zz}}\sqrt{1 + 2e'_{xx}}}. \tag{5.41}$$

and the ratio of the volume change:

$$\varepsilon'_v = 1 - \left(1 + \varepsilon'_{xx}\right)\left(1 + \varepsilon'_{yy}\right)\left(1 + \varepsilon'_{zz}\right) = 1 - \sqrt{1 - 2e'_{xx}}\sqrt{1 - 2e'_{yy}}\sqrt{1 - 2e'_{zz}}. \tag{5.42}$$

It can be written:

$$1 - \varepsilon'_v = \frac{\partial(x, y, z)}{\partial(x', y', z')} = \begin{vmatrix} 1 - \frac{\partial u}{\partial x'} & -\frac{\partial u}{\partial y'} & -\frac{\partial u}{\partial z'} \\ -\frac{\partial v}{\partial x'} & 1 - \frac{\partial v}{\partial y'} & -\frac{\partial v}{\partial z'} \\ -\frac{\partial w}{\partial x'} & -\frac{\partial w}{\partial y'} & -\frac{\partial w}{\partial z'} \end{vmatrix}. \tag{5.43}$$

5.1.4 *Infinitesimal Deformation*

If only small deformations of a body are considered, it is admitted that the strain may be neglected in relation to the unity. The product of two strains may also be neglected in relation to any other strain. Let us consider that we also have the hypothesis of small rigid body rotations, namely the components of the rigid local rotation vector are negligible relative to the unity. All of these hypotheses correspond to a linear geometric theory (previous assumptions are equivalent to the hypothesis of neglecting the motion gradient relative to the unity). In this case it can be shown that if $f(x, y, z, t)$ or $f(x', y', z', t)$ is a certain function of class C^1, expressed using the Lagrangian or Eulerian variables, there is the relation:

$$\frac{\partial f}{\partial x} = \frac{\partial f}{\partial x'}. \tag{5.44}$$

It follows that in the linear theory the same results are obtained indifferent if we use Lagrangian coordinates or Eulerian coordinates.

Thus, it can be written in the case of the linear theory:

$$\begin{aligned} \varepsilon_{xx} &= e_{xx}, \quad \varepsilon_{yy} = e_{yy}, \quad \varepsilon_{zz} = e_{zz}, \\ \gamma_{yz} &= e_{yz}, \quad \gamma_{zx} = e_{zx}, \quad \gamma_{xy} = e_{zx}. \end{aligned} \tag{5.45}$$

in Lagrangian coordinates, or

$$\begin{aligned} \varepsilon'_{xx} &= e'_{xx}, \quad \varepsilon'_{yy} = e'_{yy}, \quad \varepsilon'_{zz} = e'_{zz}, \\ \gamma'_{yz} &= e'_{yz}, \quad \gamma'_{zx} = e'_{zx}, \quad \gamma'_{xy} = e'_{zx}, \end{aligned} \tag{5.46}$$

in the Eulerian coordinates.

The strain tensors in both cases become identical. It is simply called the strains tensor:

$$[T_\varepsilon] = \begin{bmatrix} \varepsilon_{xx} & \frac{1}{2}\gamma_{xy} & \frac{1}{2}\gamma_{xz} \\ \frac{1}{2}\gamma_{xy} & \varepsilon_{yy} & \frac{1}{2}\gamma_{yz} \\ \frac{1}{2}\gamma_{xz} & \frac{1}{2}\gamma_{yz} & \varepsilon_{zz} \end{bmatrix}. \tag{5.47}$$

The strain tensor is therefore a symmetrical tensor. In the case of small deformations, neglecting small expressions of second order, the following approximations are obtained:

$$\varepsilon_{xx} = \frac{\partial u_x}{\partial x}, \quad \varepsilon_{yy} = \frac{\partial u_y}{\partial y}, \quad \varepsilon_{zz} = \frac{\partial u_z}{\partial z},$$

$$\begin{aligned} \gamma_{xy} = \gamma_{yx} &= \frac{\partial u_x}{\partial y} + \frac{\partial u_y}{\partial x}, \\ \gamma_{yz} = \gamma_{zy} &= \frac{\partial u_y}{\partial z} + \frac{\partial u_z}{\partial y}, \\ \gamma_{zx} = \gamma_{xz} &= \frac{\partial u_z}{\partial x} + \frac{\partial u_x}{\partial z}. \end{aligned} \tag{5.48}$$

5.1.5 Eigenvalues and Eigenvectors

The strain (5.24) in regards to a normal direction $\overrightarrow{n}$, becomes for small deformations:

$$\begin{aligned} \varepsilon_n &= \varepsilon_{xx}\cos^2(n,x) + \varepsilon_{yy}\cos^2(n,y) + \varepsilon_{zz}\cos^2(n,z) \\ &\quad + \gamma_{yz}\cos(n,y)\cos(n,z) + \gamma_{zx}\cos(n,z)\cos(n,x) + \gamma_{xy}\cos(n,x)\cos(n,y). \end{aligned} \tag{5.49}$$

Extremal strains are obtained by applying the Lagrange's multiplier method. A homogeneous system of linear equations will result:

$$([T_\varepsilon] - \varepsilon[E])\{v\} = 0, \tag{5.50}$$

which has solutions if and only if:

$$\det([T_\varepsilon] - \varepsilon[E]) = 0, \tag{5.51}$$

or

$$\begin{vmatrix} \varepsilon_{xx} - \varepsilon & \frac{1}{2}\gamma_{xy} & \frac{1}{2}\gamma_{xz} \\ \frac{1}{2}\gamma_{xy} & \varepsilon_{yy} - \varepsilon & \frac{1}{2}\gamma_{yz} \\ \frac{1}{2}\gamma_{xz} & \frac{1}{2}\gamma_{yz} & \varepsilon_{zz} - \varepsilon \end{vmatrix} = 0, \tag{5.52}$$

which is a third-order degree polynomial in ε:

$$\varepsilon^3 - I_1\varepsilon^2 + I_2\varepsilon - I_3 = 0. \tag{5.53}$$

The roots of this equation are always real. The coefficients are the invariants:

$$I_1 = \varepsilon_{xx} + \varepsilon_{yy} + \varepsilon_{zz}, \tag{5.54}$$

$$\begin{aligned} I_2 &= \begin{vmatrix} \varepsilon_{xx} & \frac{1}{2}\gamma_{xy} \\ \frac{1}{2}\gamma_{xy} & \varepsilon_{yy} \end{vmatrix} + \begin{vmatrix} \varepsilon_{yy} & \frac{1}{2}\gamma_{yz} \\ \frac{1}{2}\gamma_{yz} & \varepsilon_{zz} \end{vmatrix} + \begin{vmatrix} \varepsilon_{zz} & \frac{1}{2}\gamma_{zx} \\ \frac{1}{2}\gamma_{zx} & \varepsilon_{xx} \end{vmatrix} \\ &= \varepsilon_{xx}\varepsilon_{yy} + \varepsilon_{yy}\varepsilon_{zz} + \varepsilon_{zz}\varepsilon_{xx} - \frac{1}{4}\gamma_{xy}^2 - \frac{1}{4}\gamma_{yz}^2 - \frac{1}{4}\gamma_{zx}^2, \end{aligned} \tag{5.55}$$

$$I_3 = \begin{vmatrix} \varepsilon_{xx} & \frac{1}{2}\gamma_{xy} & \frac{1}{2}\gamma_{xz} \\ \frac{1}{2}\gamma_{xy} & \varepsilon_{yy} & \frac{1}{2}\gamma_{yz} \\ \frac{1}{2}\gamma_{xz} & \frac{1}{2}\gamma_{yz} & \varepsilon_{zz} \end{vmatrix}. \tag{5.56}$$

The directions in which these strains are extreme are given by the linear system (5.50), written as:

$$\begin{vmatrix} \varepsilon_{xx} - \varepsilon & \frac{1}{2}\gamma_{xy} & \frac{1}{2}\gamma_{xz} \\ \frac{1}{2}\gamma_{xy} & \varepsilon_{yy} - \varepsilon & \frac{1}{2}\gamma_{yz} \\ \frac{1}{2}\gamma_{xz} & \frac{1}{2}\gamma_{yz} & \varepsilon_{zz} - \varepsilon \end{vmatrix} \begin{Bmatrix} \alpha \\ \beta \\ \gamma \end{Bmatrix} = \begin{Bmatrix} 0 \\ 0 \\ 0 \end{Bmatrix}. \tag{5.57}$$

The normalization of the eigenvector can be made using the condition:

$$\alpha^2 + \beta^2 + \gamma^2 = 1. \tag{5.58}$$

5.1.6 The Physical Significance of the Components of the Strain Tensor

To see the physical significance of the quantities ε_{xx}, ε_{yy}, ε_{zz}, we consider the vector $\{\mathrm{d}r\} = \begin{bmatrix} 1 & 0 & 0 \end{bmatrix}^T \mathrm{d}x$ which, after deformation, becomes:

$$\{\mathrm{d}r'\} = ([E] + [\partial]) \begin{Bmatrix} 1 \\ 0 \\ 0 \end{Bmatrix} \mathrm{d}x. \tag{5.59}$$

The square of the module of this vector is obtained, in the approximation of small deformations, by:

$$\begin{aligned} \{\mathrm{d}r'\}^2 &= \{\mathrm{d}r'\}^T \{\mathrm{d}r'\} = [\,1\;0\;0\,]([E] + 2[\varepsilon]) \begin{Bmatrix} 1 \\ 0 \\ 0 \end{Bmatrix} (\mathrm{d}x)^2 \\ &= (1 + 2\varepsilon_{xx})(\mathrm{d}x)^2. \end{aligned} \tag{5.60}$$

It follows that $\mathrm{d}r' = \sqrt{1 + 2\varepsilon_{xx}}\mathrm{d}x \approx (1 + \varepsilon_{xx})\mathrm{d}x$, where the approximation was made considering a Taylor's expansion of the radical. In this case the strain of the vector $\mathrm{d}r$ is:

$$\frac{\mathrm{dr'} - \mathrm{dr}}{\mathrm{dr}} = \frac{\sqrt{1 + 2\varepsilon_{xx}}\mathrm{dx} - \mathrm{dx}}{\mathrm{dx}} \cong \varepsilon_{\mathrm{XX}}, \tag{5.61}$$

Because $(\mathrm{d}x)^2 = (\mathrm{d}r)^2$. It follows that the element ε_{xx} of the strain tensor is the strain of an infinitesimal segment oriented along the Ox-axis. Similar meanings have ε_{xx} and ε_{zz}.

To see the physical significance of the quantities ε_{xy}, ε_{yz} and ε_{xz} take the vectors $\{\mathrm{d}r_1\} = [\,1\;0\;0\,]^T \mathrm{d}x_1$ and $\{\mathrm{d}r_2\} = [\,1\;0\;0\,]^T \mathrm{d}x_2$ and observe the variation of the angle between the two vectors during the deformation:

$$\left\{\mathrm{d}r'_1\right\} = ([E] + [\partial])\{\mathrm{d}r_1\}; \quad \left\{\mathrm{d}r'_2\right\} = ([E] + [\partial])\{\mathrm{d}r_2\}. \tag{5.62}$$

The dot product between the two vectors will be, assuming small deformations, and taking into account the previously written relationships:

$$\left\{\mathrm{d}r'_1\right\}^{\mathrm{T}} \left\{\mathrm{d}r'_2\right\} = \{\mathrm{d}r_1\}^{\mathrm{T}}([E] + 2[\varepsilon]^T)\{\mathrm{d}r_2\} = 2\varepsilon_{xy}\mathrm{d}x_1\mathrm{d}x_2 \tag{5.63}$$

but:

$$\{dr_1'\}^T\{dr_2'\} = dx_1'dx_2'\cos(d\bar{r}_1', d\bar{r}_2').$$
$$\{dr\}^T\{dr\} = \{dr'\}^T([E] - [\partial]^T)([E] - [\partial])\{dr'\} \tag{5.64}$$

If we consider that we have $dx_1'dx_2' \cong dx_1dx_2$ it results, for small deformations, the relationship:

$$2\varepsilon_{xy} = \cos(d\bar{r}_1', d\bar{r}_2') = \sin\left(\frac{\pi}{2} - < (d\bar{r}_1', d\bar{r}_2')\right) = \sin\gamma \cong \gamma, \tag{5.65}$$

where γ is the variation of the right angle made by the two vectors considered (i.e. by the axes Ox and Oy). We have:

$$\gamma = \gamma_1 + \gamma_2, \tag{5.66}$$

where γ_1 is the angle of the vectors $d\bar{r}_1$ and $d\bar{r}_1'$ and γ_2 is the angle made by $d\bar{r}_2$ and $d\bar{r}_2'$. Similar meanings have ε_{yz} and ε_{xz}.

5.1.7 Transformation Induced by the Strain Tensor

Let us study what becomes an infinitesimal sphere, centered on point P, after deformation. We have:

$$\{dr'\} = ([E] + [\partial])^{-1}\{dr\}, \tag{5.67}$$

so:

$$\{dr\} = ([E] + [\partial])^{-1}\{dr'\}. \tag{5.68}$$

In the case of small deformations, we can approximate:

$$([E] + [\partial])^{-1} \cong [E] - [\partial]. \tag{5.69}$$

Let us now consider the infinitesimal sphere:

$$(dr)^2 = (dx)^2 + (dy)^2 + (dz)^2. \tag{5.70}$$

Taking into account the transformation presented, it can be written:

$$\begin{aligned}(dr)^2 &= \{dr\}^T\{dr\} = \{dr'\}^T([E] - [\partial]^T)([E] - [\partial])\{dr'\} \\ &= \{dr'\}^T([E] - [\partial] - [\partial]^T + [\partial]^T[\partial])\{dr'\} \\ &= \{dr'\}^T([E] - [\partial] - [\partial]^T)\{dr'\} = \{dr'\}^T([E] - 2[\varepsilon]^T)\{dr'\},\end{aligned} \tag{5.71}$$

which represents the equation of an ellipsoid. The directions of the new coordinate system against which the equation of the ellipsoid can be brought into the canonical form are the matrix's eigenvectors and the ellipsoid semi-axes are given by the eigenvalues of this matrix.

5.1.8 Local Rigid Rotation

Consider now that we are in the case of small deformations and the relations (5.48) are valid. We introduce the skew symmetric matrix:

$$[T_\omega] = \begin{bmatrix} 0 & \omega_{xy} & -\omega_{zx} \\ -\omega_{xy} & 0 & \omega_{yz} \\ \omega_{zx} & -\omega_{yz} & 0 \end{bmatrix}, \tag{5.72}$$

with three distinct components:

$$\omega_{yz} = \frac{1}{2}\left(\frac{\partial u_z}{\partial y} - \frac{\partial u_y}{\partial z}\right), \quad \omega_{zx} = \frac{1}{2}\left(\frac{\partial u_x}{\partial z} - \frac{\partial u_z}{\partial x}\right), \quad \omega_{xy} = \frac{1}{2}\left(\frac{\partial u_y}{\partial x} - \frac{\partial u_x}{\partial y}\right). \tag{5.73}$$

The vector

$$\{\omega\}^T = \begin{bmatrix} \omega_{yz} \; \omega_{zx} \; \omega_{xy} \end{bmatrix},$$

is called the local rigid rotation vector.

The differential of the displacement $\bar{u}$ is:

$$\{du\} = \begin{Bmatrix} du_x \\ du_y \\ du_z \end{Bmatrix} = \begin{bmatrix} \frac{\partial u_x}{\partial x} & \frac{\partial u_x}{\partial y} & \frac{\partial u_x}{\partial z} \\ \frac{\partial u_y}{\partial x} & \frac{\partial u_y}{\partial y} & \frac{\partial u_y}{\partial z} \\ \frac{\partial u_z}{\partial x} & \frac{\partial u_z}{\partial y} & \frac{\partial u_z}{\partial z} \end{bmatrix} \begin{Bmatrix} dx \\ dy \\ dz \end{Bmatrix}$$
$$= \begin{bmatrix} \alpha_{xx} & \alpha_{xy} & \alpha_{xz} \\ \alpha_{yx} & \alpha_{yy} & \alpha_{yz} \\ \alpha_{zx} & \alpha_{zy} & \alpha_{zz} \end{bmatrix} \begin{Bmatrix} dx \\ dy \\ dz \end{Bmatrix} = [T_\alpha]\{dr\}. \tag{5.74}$$

The matrix:

$$[T_\alpha] = \begin{bmatrix} \alpha_{xx} & \alpha_{xy} & \alpha_{xz} \\ \alpha_{yx} & \alpha_{yy} & \alpha_{yz} \\ \alpha_{zx} & \alpha_{zy} & \alpha_{zz} \end{bmatrix},$$

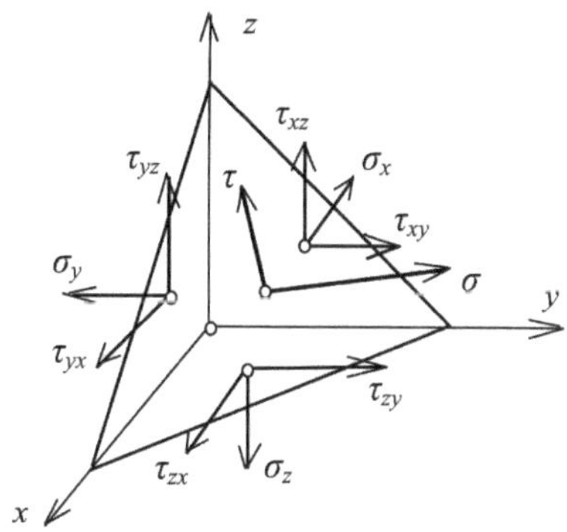

Fig. 5.3 Components of the stress tensor

is called the gradient of the displacement $\bar{u}$.

It can be written:

$$[T_\alpha] = [T_\varepsilon] + [T_\omega], \tag{5.75}$$

that is, the displacement gradient tensor can be decomposed into two tensors, one symmetrical $[T_\varepsilon]$ (the strain tensor) and one skew symmetric $[T_\omega]$ (the rigid local rotation tensor).

The rigid displacement is given by:

$$\mathrm{d}\overline{u} = \overline{\omega} \times \mathrm{d}\overline{r}, \tag{5.76}$$

$$\{\mathrm{d}u\} = \begin{Bmatrix} \mathrm{d}u_x \\ \mathrm{d}u_y \\ \mathrm{d}u_z \end{Bmatrix} = \begin{vmatrix} \bar{\imath} & \bar{\jmath} & \bar{k} \\ \omega_{yz} & \omega_{zx} & \omega_{xy} \\ \mathrm{d}x & \mathrm{d}y & \mathrm{d}z \end{vmatrix} = \begin{Bmatrix} -\omega_{xy}\mathrm{d}y + \omega_{zx}\mathrm{d}z \\ -\omega_{yz}\mathrm{d}z + \omega_{xy}\mathrm{d}x \\ -\omega_{zx}\mathrm{d}x + \omega_{yz}\mathrm{d}y \end{Bmatrix}. \tag{5.77}$$

5.2 Stress Tensor

5.2.1 Stress State in a Point

Let us consider an elementary volume, in the form of a tetrahedron, around a point P. On the inclined surface of the tetrahedron we consider a stress $\overrightarrow{p}$ that has the components p_{nx}, p_{ny}, p_{nz} in relation to the axes of the coordinate system. We decompose the stress into two components, one, σ, in the normal direction to the plane and to other, τ, in the tangential direction (see Fig. 5.3).

$$p^2 = \sigma^2 + \tau^2. \tag{5.78}$$

Let l, m and n be the direction cosines of the normal to the plan where $\overrightarrow{p}$ acts. The equations of equilibrium of the elementary tetrahedron lead to the expressions:

$$\begin{aligned} p_{nx} &= \sigma_x l + \tau_{yx} m + \tau_{zx} n, \\ p_{ny} &= \tau_{xy} l + \sigma_y m + \tau_{zy} n, \\ p_{nz} &= \tau_{xz} l + \tau_{yz} m + \sigma_z n, \end{aligned} \tag{5.79}$$

obtained from the balance of the forces acting on the faces of the tetraedron and:

$$\tau_{xy} = \tau_{yx}, \quad \tau_{yz} = \tau_{zy}, \quad \tau_{zx} = \tau_{xz}, \tag{5.80}$$

obtained from the balance of moments.

Grouped, one can write:

$$\begin{Bmatrix} p_{nx} \\ p_{ny} \\ p_{nz} \end{Bmatrix} = \begin{bmatrix} \sigma_{xx} & \tau_{xy} & \tau_{xz} \\ \tau_{yx} & \sigma_{yy} & \tau_{yz} \\ \tau_{zx} & \tau_{zy} & \sigma_{zz} \end{bmatrix} \begin{Bmatrix} l \\ m \\ n \end{Bmatrix}, \tag{5.80'}$$

or

$$\{p\} = [T_\sigma]\{\eta\}, \tag{5.80''}$$

where were used the notations:

$$\{p\} = \begin{Bmatrix} p_{nx} \\ p_{ny} \\ p_{nz} \end{Bmatrix}, \quad [T_\sigma] = \begin{bmatrix} \sigma_{xx} & \tau_{xy} & \tau_{xz} \\ & \sigma_{yy} & \tau_{yz} \\ SIM & & \sigma_{zz} \end{bmatrix}, \quad \{\eta\} = \begin{Bmatrix} l \\ m \\ n \end{Bmatrix}.$$

The tensor $[T_\sigma]$ is called the stress tensor.

5.2.2 *Transformation of the Stress Tensor to Axis Rotation*

Let us consider a $Ox'y'z'$ coordinate system of the same origin as the $Oxyz$ system and which has undergone a spatial (three-dimensional) rotation. The rotation matrix is defined by:

$$[R] = [\{e_1\}\{e_2\}\{e_3\}] = \begin{bmatrix} l_1 & l_2 & l_3 \\ m_1 & m_2 & m_3 \\ n_1 & n_2 & n_3 \end{bmatrix}, \tag{5.81}$$

where $\vec{e}_1, \vec{e}_2, \vec{e}_3$ are the unit vectors of the Ox', Oy', Oz' axes. In the $Oxyz$ system, we have the normal vector $\vec{n}$ with the components l, m, n and the normal stress $\vec{p}$ with the components p_{nx}, p_{ny}, p_{nz} and in the $Ox'y'z$ system, the normal and normal stress components are: l', m', n' and, respectively p'_{nx}, p'_{ny}, p'_{nz}.

The transformation relationships of the two vectors in the transition from one coordinate system to the other are:

$$\begin{Bmatrix} l \\ m \\ n \end{Bmatrix} = [R] \begin{Bmatrix} l' \\ m' \\ n' \end{Bmatrix} \quad \text{and} \quad \begin{Bmatrix} p_{nx} \\ p_{ny} \\ p_{nz} \end{Bmatrix} = [R] \begin{Bmatrix} p'_{nx} \\ p'_{ny} \\ p'_{nz} \end{Bmatrix}. \tag{5.82}$$

If $\{p\}$ and $\{n\}$ are replaced in relation (5.80″), one obtains:

$$[R]\{p'\} = [T_\sigma][R]\{\eta\}, \tag{5.83}$$

which, premultiplied with $[R]^T$, offers:

$$\{p'\} = [R]^T[T_\sigma][R]\{\eta'\}. \tag{5.84}$$

It follows:

$$[T'_\sigma] = [R]^T[T_\sigma][R], \tag{5.84'}$$

which represents the formula of changing the components of the stress tensors when changing the coordinate system.

5.2.3 *Normal Stress Corresponding to an Arbitrary Direction*

Let us consider the stress vector at a surface given by the formula (5.80″). The projection of this vector following the normal direction to the surface is:

$$\begin{aligned} \sigma_n &= \{\eta\}^T\{p\} = \{\eta\}^T[T]\{\eta\} \\ &= [l\ m\ n] \begin{bmatrix} \sigma_{xx} & \tau_{xy} & \tau_{xz} \\ & \sigma_{yy} & \tau_{yz} \\ SIM & & \sigma_{zz} \end{bmatrix} \begin{Bmatrix} l \\ m \\ n \end{Bmatrix} \\ &= \sigma_{xx}l^2 + \sigma_{yy}m^2 + \sigma_{zz}n^2 + 2\tau_{xy}lm + 2\tau_{yz}mn + 2\tau_{zx}nl. \end{aligned} \tag{5.85}$$

5.2.4 Extremal Conditions for Normal Stress

We propose to determine the directions for which the normal tension has extremal values. Thus, we want to minimize the function $\sigma_n(l, m, n)$ for which the variables l, m, n have to fulfill the normality condition: $l^2 + m^2 + n^2 = 1$.

To determine the extremal values we use the Lagrange multipliers method by choosing the objective function:

$$L(l, m, n) = \sigma_n(l, m, n) - \lambda\big(l^2 + m^2 + n^2 - 1\big). \tag{5.86}$$

The extremal conditions for this function using the Lagrange's multiplier method are:

$$\frac{\partial L}{\partial l} = 0, \quad \frac{\partial L}{\partial m} = 0, \quad \frac{\partial L}{\partial n} = 0, \quad \frac{\partial L}{\partial \lambda} = 0. \tag{5.87}$$

The first three relationships lead to the system:

$$\begin{aligned} (\sigma_{xx} - \lambda)l + \tau_{xy}m + \tau_{xz}n &= 0, \\ \tau_{xy}l + \big(\sigma_{yy} - \lambda\big)m + \tau_{yz}n &= 0, \\ \tau_{xz}l + \tau_{xy}m + \big(\sigma_{yy} - \lambda\big)n &= 0, \end{aligned} \tag{5.88}$$

and the fourth to

$$l^2 + m^2 + n^2 = 1. \tag{5.89}$$

The first three relationships represent a homogeneous linear system that can be written as:

$$([T_\sigma] - \lambda[E])\{\eta\} = \{0\}. \tag{5.90}$$

It follows that the directions for which the normal stress reaches their extremal values are their eigenvectors for the stress matrix and the values that the stresses are taking in this case are the eigenvalues of this matrix. The obtained homogeneous system obtained has solutions if and only if the following condition is satisfied:

$$\det([T_\sigma] - \lambda[E]) = \{0\}, \tag{5.91}$$

namely:

$$\begin{vmatrix} \sigma_{xx} - \lambda & \tau_{xy} & \tau_{xz} \\ \tau_{xy} & \sigma_{yy} - \lambda & \tau_{yz} \\ \tau_{xz} & \tau_{yz} & \sigma_{zz} - \lambda \end{vmatrix} = 0. \tag{5.92}$$

Since the stress matrix is symmetric, the written equation has real roots. The characteristic equation is a third order polynomial of the form:

$$\lambda^3 - I_1\lambda^2 + I_2\lambda - I_3 = 0, \tag{5.93}$$

or, if the eigenvalue λ represents a stress, in the form of:

$$\sigma^3 - I_1\sigma^2 + I_2\sigma - I_3 = 0, \tag{5.94}$$

where I_1, I_2, I_3 are the invariants of the stress tensor:

$$I_1 = \sigma_{xx} + \sigma_{yy} + \sigma_{zz}, \tag{5.94$'$}$$

$$\begin{aligned} I_2 &= \begin{vmatrix} \sigma_{xx} & \tau_{xy} \\ \tau_{xy} & \sigma_{yy} \end{vmatrix} + \begin{vmatrix} \sigma_{yy} & \tau_{yz} \\ \tau_{yz} & \sigma_{zz} \end{vmatrix} + \begin{vmatrix} \sigma_{xx} & \tau_{xz} \\ \tau_{xz} & \sigma_{zz} \end{vmatrix} \\ &= \sigma_{xx}\sigma_{yy} + \sigma_{yy}\sigma_{zz} + \sigma_{zz}\sigma_{xx} - \tau_{xy}^2 - \tau_{yz}^2 - \tau_{zx}^2, \end{aligned} \tag{5.95}$$

$$I_3 = \begin{vmatrix} \sigma_{xx} & \tau_{xy} & \tau_{xz} \\ \tau_{xy} & \sigma_{yy} & \tau_{yz} \\ \tau_{xz} & \tau_{yz} & \sigma_{zz} \end{vmatrix}. \tag{5.96}$$

The quantities I_1, I_2, I_3 are independent of the chosen coordinate system. The first and the second invariant, respectively are called the linear invariant and the quadratic invariant, have particular physical significance in the theory of plasticity, explained below. The roots of the characteristic equation are the principal normal stresses $\sigma_1, \sigma_2, \sigma_3$.

The characteristic equation can be solved algebraically, with the formula of Cardan-Tartaglia, or by trigonometric or numerical methods. We present the trigonometric method.

We introduce the unknown θ and substitute:

$$\sigma = r\cos\theta - \frac{I_1}{3}. \tag{5.97}$$

With this variable change the characteristic equation becomes:

$$\cos^3\theta - \frac{\left(I_1^2 - 3I_2\right)}{3r^2}\cos\theta - \frac{2I_1^3 - 9I_1I_2 + 27I_3}{27r^3} = 0. \tag{5.98}$$

Taking into account the trigonometric identity:

$$4\cos^3\theta - 3\cos\theta = \cos 3\theta, \tag{5.99}$$

written as:

$$\cos^3\theta - \frac{3}{4}\cos\theta - \frac{1}{4}\cos 3\theta = 0, \tag{5.100}$$

and comparison with the characteristic equation, it follows by identification:

$$r = \frac{2}{3}\left(I_1^2 - 3I_2\right)^{\frac{1}{2}} \quad \text{and} \quad \cos 3\theta = \frac{2I_1^3 - 9I_1I_2 + 27I_3}{2\left(I_1^2 - 3I_2\right)^{\frac{3}{2}}}, \tag{5.101}$$

from where:

$$\theta_k = \frac{\pm\theta_o + 2k\pi}{3} \quad \text{and} \quad \theta_o = \arccos\frac{2I_1^3 - 9I_1I_2 + 27I_3}{2\left(I_1^2 - 3I_2\right)^{\frac{3}{2}}}. \tag{5.102}$$

The expression of the three invariants according to the principal stresses are:

$$I_1 = \sigma_1 + \sigma_2 + \sigma_3, \tag{5.103}$$

$$I_2 = \sigma_1\sigma_2 + \sigma_2\sigma_3 + \sigma_3\sigma_1, \tag{5.104}$$

$$I_3 = \sigma_1\sigma_2\sigma_3. \tag{5.105}$$

We further demonstrate that the eigenvectors of the stress tensor are orthogonal. If $\{\eta_i\}$ it is an eigenvector of the matrix $[T_\sigma]$ we have:

$$[T_\sigma]\{\eta_i\} = \lambda_i\{\eta_i\}. \tag{5.106}$$

Multiplying with $\{\eta_j\}^T$, one obtains:

$$\{\eta_j\}^T[T_\sigma]\{\eta_i\} = \lambda_i\{\eta_j\}^T\{\eta_i\}. \tag{5.107}$$

Similarly, there is also obtained:

$$\{\eta_i\}^T[T_\sigma]\{\eta_j\} = \lambda_j\{\eta_i\}^T\{\eta_j\}. \tag{5.108}$$

But since $[T_\sigma]$ is symmetric, one obtains:

$$\{\eta_i\}^T[T_\sigma]\{\eta_j\} = \{\eta_j\}^T[T_\sigma]\{\eta_i\} \quad \text{and} \quad \{\eta_i\}^T\{\eta_j\} = \{\eta_j\}^T\{\eta_i\}.$$

Substracting relations (5.107) and (5.108) gives:

$$0 = \left(\lambda_i - \lambda_j\right)\{\eta_i\}^T\left\{\eta_j\right\}, \tag{5.109}$$

where, as a general rule $\lambda_i \neq \lambda_j$, it is obtained:

$$\{\eta_i\}^T\left\{\eta_j\right\} = 0, \tag{5.110}$$

that is the eigenvectors of the stress tensor are orthogonal. It follows that the three main directions of inertia can define an orthogonal reference. We can write the relationships:

$$\{\eta_i\}^T[T_\sigma]\left\{\eta_j\right\} = \delta_{ij}\lambda_i, \tag{5.111}$$

where δ_{ij} is the Kronecker's symbol. Introducing $[R]$ as the rotation that makes the transition from the coordinate system in which the stress tensor is expressed to the coordinate system defined by its eigenvectors:

$$[R] = [\{\eta_1\}\{\eta_2\}\{\eta_3\}]. \tag{5.112}$$

In this new coordinate system the stress tensor is expressed by:

$$\begin{aligned}
[T_\sigma]' &= [R]^T[T_\sigma][R] \\
&= \begin{bmatrix} \{\eta_1\}^T \\ \{\eta_2\}^T \\ \{\eta_3\}^T \end{bmatrix} [T_\sigma][\{\eta_1\}\{\eta_2\}\{\eta_3\}] \\
&= \begin{bmatrix} \{\eta_1\}^T[T]\{\eta_1\} & \{\eta_1\}^T[T]\{\eta_2\} & \{\eta_1\}^T[T]\{\eta_3\} \\ \{\eta_2\}^T[T]\{\eta_1\} & \{\eta_2\}^T[T]\{\eta_2\} & \{\eta_2\}^T[T]\{\eta_3\} \\ \{\eta_3\}^T[T]\{\eta_1\} & \{\eta_3\}^T[T]\{\eta_2\} & \{\eta_3\}^T[T]\{\eta_3\} \end{bmatrix} \\
&= \begin{bmatrix} \sigma_1 & 0 & 0 \\ 0 & \sigma_2 & 0 \\ 0 & 0 & \sigma_2 \end{bmatrix},
\end{aligned} \tag{5.113}$$

so that the stress tensor takes the diagonal form.

5.2.5 *Invariants of the Reduced Stress*

There is another form of writing the characteristic equation, useful in some applications. Thus, if the substitution is made:

$$\sigma = \sigma' + \frac{I_1}{3}, \tag{5.114}$$

the characteristic equation becomes:

$$\left(\sigma'\right)^3 - J_1\left(\sigma'\right)^2 + J_2\sigma' - J_3 = 0, \tag{5.115}$$

where the following notations were used:

$$J_1 = 0, \quad J_2 = \frac{I_1 - 3I_2}{3}, \quad J_3 = \frac{2I_1^3 - 9I_1I_2 + 27I_3}{27}. \tag{5.116}$$

The quantities J_1, J_2, J_3 are called the first, second and third invariants of the reduced stress and are of particular importance when considering the flow of metals. If σ_1', σ_2', σ_3' represent the roots of the transformed characteristic equation, it can be written:

$$J_1 = \sigma_1', \sigma_2', \sigma_3' = 0, \tag{5.117}$$

from where:

$$\begin{aligned} J_2 &= \sigma_1'\sigma_2' + \sigma_2'\sigma_3' + \sigma_3'\sigma_1' \\ &= \frac{1}{2}\left[\left(\sigma_1' + \sigma_2' + \sigma_3'\right)^2 - \left(\sigma_1'\right)^2 - \left(\sigma_2'\right)^2 - \left(\sigma_3'\right)^2\right] \\ &= -\frac{1}{2}\left(\left(\sigma_1'\right)^2 + \left(\sigma_2'\right)^2 + \left(\sigma_3'\right)^2\right); \\ J_3 &= \sigma_1'\sigma_2'\sigma_3' = \frac{1}{3}\left(\left(\sigma_1'\right)^3 + \left(\sigma_2'\right)^3 + \left(\sigma_3'\right)^3\right). \end{aligned} \tag{5.118}$$

5.2.6 *Conic of Normal Stress*

If we are in a plane, the stress tensor has the form:

$$[\sigma] = \begin{bmatrix} \sigma_x & \tau_{xy} \\ \tau_{xy} & \sigma_y \end{bmatrix}. \tag{5.119}$$

Let us now consider in the *Oxy* plane a direction defined by the unit vector $\vec{\eta}\,(\cos\alpha,\ \sin\alpha)$. The value of normal stress in this direction is given by:

$$\sigma_n = \{\eta\}^T[\sigma]\{\eta\}. \tag{5.120}$$

To represent this stress, let us take in this direction a segment originating in O and having the size d, chosen so $\sigma_n d^2 = k^2$ that:

$$d = \frac{|k|}{\sqrt{\sigma_n}}. \tag{5.121}$$

Consider the vector $\overrightarrow{r}$ given by

$$\overrightarrow{r} = d\overrightarrow{\eta} = d\cos\alpha\,\overrightarrow{i} + d\sin\alpha\,\overrightarrow{j} = x\,\overrightarrow{i} + y\,\overrightarrow{j}.$$

Then multiplying the previously written relationship with d^2, one obtains:

$$\sigma_n d^2 = d\{\eta\}^T[\sigma]d\{\eta\} = \{r\}^T[\sigma]\{r\} = k^2, \tag{5.122}$$

or

$$\begin{bmatrix} x & y \end{bmatrix}\begin{bmatrix} \sigma_x & \tau_{xy} \\ \tau_{xy} & \sigma_y \end{bmatrix}\begin{Bmatrix} x \\ y \end{Bmatrix} = k^2. \tag{5.123}$$

5.2.7 *Quadric of Normal Stress*

Let us consider the vector $\overrightarrow{MQ}$ that has the size:

$$MQ = \frac{1}{\sqrt{\pm\sigma_n}}, \tag{5.124}$$

with the components:

$$x = \frac{\cos(\eta, x)}{\sqrt{\pm\sigma_n}} = \frac{l}{\sqrt{\pm\sigma_n}}, \quad y = \frac{\cos(\eta, y)}{\sqrt{\pm\sigma_n}} = \frac{m}{\sqrt{\pm\sigma_n}}, \quad z = \frac{\cos(\eta, z)}{\sqrt{\pm\sigma_n}} = \frac{n}{\sqrt{\pm\sigma_n}}, \tag{5.125}$$

pointing in the direction of the outer normal $\overrightarrow{\eta}$ to the considered area element.

Using (5.85), we find that the set of all points Q in space is a quadratic (M is the origin of the coordinate system):

$$\sigma_{xx}x^2 + \sigma_{yy}y^2 + \sigma_{zz}z^2 + 2\tau_{xy}xy + 2\tau_{yz}myz + 2\tau_{zx}zx = \pm 1. \tag{5.126}$$

The plus sign is taken if we have tension and the minus sign is taken if we have compression so that the surface obtained is real. This surface is called the quadric of the normal stress (quadric of Cauchy).

There are therefore three perpendicular directions after which the extremal normal stresses $\sigma_1 \geq \sigma_2 \geq \sigma_3$ develop and for which the tangential stresses are null. Generally speaking, these three principal directions do not coincide with the principal directions determined for the strain tensor. The principal directions for the two tensors will coincide in the particular case of isotropic linear elastic bodies (lacking the tangential stresses missing and the corresponding shear strains and vice versa) and in the case of anisotropic linear elastic bodies.

5.2.8 Constitutive Equations

The geometric and mechanical aspect must be completed by a physical, experimental aspect. A constitutive law is introduced, that means a relation between the stresses and the strains, considered valid the assumptions made so far. For this, the generalized Hooke's law (coming from the one-dimensional case) is used. Between the stress tensor and the strain tensor there is a linear relation of the form:

$$[T_\varepsilon] = [H][T_\sigma], \tag{5.127}$$

where $[H]$ is the Hooke's tensor (Teodorescu 1972):

$$[H] = \begin{bmatrix} \frac{1}{E_x} & -\frac{\mu_{xy}}{E_y} & -\frac{\mu_{xz}}{E_z} & \frac{\eta_{x,yz}}{G_{yz}} & \frac{\eta_{x,zx}}{G_{zx}} & \frac{\eta_{x,xy}}{G_{xy}} \\ -\frac{\mu_{yx}}{E_x} & \frac{1}{E_y} & -\frac{\mu_{yz}}{E_z} & \frac{\eta_{y,yz}}{G_{yz}} & \frac{\eta_{y,zx}}{G_{zx}} & \frac{\eta_{y,xy}}{G_{xy}} \\ -\frac{\mu_{zx}}{E_x} & -\frac{\mu_{zy}}{E_y} & \frac{1}{E_z} & \frac{\eta_{z,yz}}{G_{yz}} & \frac{\eta_{z,zx}}{G_{zx}} & \frac{\eta_{z,xy}}{G_{xy}} \\ \frac{\eta_{yz,x}}{E_x} & \frac{\eta_{yz,y}}{E_y} & \frac{\eta_{yz,z}}{E_z} & \frac{1}{G_{yz}} & \frac{\upsilon_{yz,zx}}{G_{zx}} & \frac{\upsilon_{yz,xy}}{G_{xy}} \\ \frac{\eta_{zx,x}}{E_x} & \frac{\eta_{zx,y}}{E_y} & \frac{\eta_{zx,z}}{E_z} & \frac{\upsilon_{zx,yz}}{G_{yz}} & \frac{1}{G_{zx}} & \frac{\upsilon_{zx,xy}}{G_{xy}} \\ \frac{\eta_{xy,x}}{E_x} & \frac{\eta_{xy,y}}{E_y} & \frac{\eta_{xy,z}}{E_z} & \frac{\upsilon_{xy,yz}}{G_{yz}} & \frac{\upsilon_{xy,zx}}{G_{zx}} & \frac{1}{G_{xy}} \end{bmatrix} \tag{5.128}$$

The matrix $[H]$ is symmetric, so of the 36 elements that define it, only 21 are independent. The symmetry conditions impose:

$$\frac{\mu_{xy}}{E_y} = \frac{\mu_{yx}}{E_x};\ \frac{\mu_{xz}}{E_z} = \frac{\mu_{zx}}{E_x};\ \frac{\mu_{yz}}{E_z} = \frac{\mu_{zy}}{E_y}, \tag{5.129}$$

$$\frac{\upsilon_{yz,zx}}{G_{zx}} = \frac{\upsilon_{zx,yz}}{G_{yz}};\ \frac{\upsilon_{yz,xy}}{G_{xy}} = \frac{\upsilon_{xy,yz}}{G_{yz}};\ \frac{\upsilon_{zx,xy}}{G_{xy}} = \frac{\upsilon_{xy,zx}}{G_{zx}}, \tag{5.130}$$

$$\frac{\eta_{x,yz}}{G_{yz}} = \frac{\eta_{yz,x}}{E_x};\ \frac{\eta_{x,zx}}{G_{zx}} = \frac{\eta_{zx,x}}{E_x};\ \frac{\eta_{x,xy}}{G_{xy}} = \frac{\eta_{xy,x}}{E_x}, \tag{5.131}$$

$$\frac{\eta_{y,yz}}{G_{yz}} = \frac{\eta_{zx,y}}{E_y};\ \frac{\eta_{y,zx}}{G_{zx}} = \frac{\eta_{zx,y}}{E_y};\ \frac{\eta_{y,xy}}{G_{xy}} = \frac{\eta_{xy,y}}{E_y}, \tag{5.132}$$

$$\frac{\eta_{z,yz}}{G_{yz}} = \frac{\eta_{yz,z}}{E_z};\ \frac{\eta_{z,zx}}{G_{zx}} = \frac{\eta_{zx,z}}{E_z};\ \frac{\eta_{z,xy}}{G_{xy}} = \frac{\eta_{xy,z}}{E_z}. \tag{5.133}$$

Particular case

We propose to see for what kind of generalized Hooke's transformation the stress tensor and strain tensor have the same principal directions. For the principal directions of the tensor, the principal tangential stresses must be null. For the principal directions of the strain tensor, the principal shear strains must be null. This happens if the relationships between stresses and deformations are:

$$\begin{aligned}
\varepsilon_x &= \frac{1}{E_x}\sigma_x - \frac{\mu_{xy}}{E_y}\sigma_y - \frac{\mu_{xz}}{E_z}\sigma_z,\\
\varepsilon_y &= -\frac{\mu_{yx}}{E_x}\sigma_x + \frac{1}{E_y}\sigma_y - \frac{\mu_{yz}}{E_z}\sigma_z,\\
\varepsilon_z &= -\frac{\mu_{zx}}{E_x}\sigma_x - \frac{\mu_{zy}}{E_y}\sigma_y + \frac{1}{E_z}\sigma_z,
\end{aligned}\tag{5.134}$$

$$\begin{aligned}
\gamma_{yz} &= \frac{1}{G_{yz}}\tau_{yz} + \frac{\nu_{yz,zx}}{G_{zx}}\tau_{zx} + \frac{\nu_{yz,xy}}{G_{xy}}\tau_{xy},\\
\gamma_{zx} &= \frac{\nu_{zx,yz}}{G_{yz}}\tau_{yz} + \frac{1}{G_{zx}}\tau_{zx} + \frac{\nu_{zx,xy}}{G_{xy}}\tau_{xy},\\
\gamma_{xy} &= \frac{\nu_{xy,yz}}{G_{yz}}\tau_{yz} + \frac{\nu_{xy,zx}}{G_{zx}}\tau_{zx} + \frac{1}{G_{xy}}\tau_{xy}.
\end{aligned}\tag{5.135}$$

In this case the material is characterized by 12 independent elastic constants.

Orthotropic material

An orthotropic material is a material that has mechanical properties different along three mutually-orthogonal twofold axes of rotational symmetry. In this case, Eqs. (5.134) and (5.135) become:

$$\begin{aligned}
\varepsilon_x &= \frac{1}{E_x}\sigma_x - \frac{\mu_{xy}}{E_y}\sigma_y - \frac{\mu_{xz}}{E_z}\sigma_z,\\
\varepsilon_y &= -\frac{\mu_{yx}}{E_x}\sigma_x + \frac{1}{E_y}\sigma_y - \frac{\mu_{yz}}{E_z}\sigma_z,\\
\varepsilon_z &= -\frac{\mu_{zx}}{E_x}\sigma_x - \frac{\mu_{zy}}{E_y}\sigma_y + \frac{1}{E_z}\sigma_z,
\end{aligned}\tag{5.136}$$

$$\gamma_{yz} = \tfrac{1}{G_{yz}}\tau_{yz},\ \gamma_{zx} = \tfrac{1}{G_{zx}}\tau_{zx},\ \gamma_{xy} = \tfrac{1}{G_{xy}}\tau_{xy}.\tag{5.137}$$

The material is characterized by nine independent parameters.

Isotropic material

In the case of the classical isotropic materials the generalized Hooke's law becomes:

$$\varepsilon_x = \frac{1}{E_x}\left[\sigma_x - \mu\left(\sigma_y + \sigma_z\right)\right],$$
$$\varepsilon_y = \frac{1}{E_x}\left[\sigma_y - \mu(\sigma_z + \sigma_x)\right],$$
$$\varepsilon_z = \frac{1}{E_z}\left[\sigma_z - \mu\left(\sigma_x + \sigma_y\right)\right]. \tag{5.138}$$

$$\gamma_{yz} = \tfrac{1}{G}\tau_{yz},\ \gamma_{zx} = \tfrac{1}{G}\tau_{zx},\ \gamma_{xy} = \tfrac{1}{G}\tau_{xy}, \tag{5.139}$$

where:

$$G = \frac{E}{2(1+\mu)}.$$

The material is characterized only by two material constants.

References

Gould, P.L.: Introduction to Linear Elasticity. Springer, Berlin (2013)
Landau, L., Lifchitz, E.: Théorie de l´élasticité. Editions Mir, Moscou (1967)
Lurie, A.I., Belyaev, A.: Theory of Elasticity (Foundations of Engineering Mechanics). Springer, Berlin (2005)
Maceri, A.: Theory of Elasticity. Springer, Berlin (2010)
Sharpe, W.N. Jr. (ed.): Springer Handbook of Experimental Solid Mechanics. Springer, Berlin (2008)
Teodorescu, P.P.: Dynamics of the Linear Elastic Bodies. Ed. Academiei, Bucharest (1970)
Teodorescu, P.P.: Dynamics of the linear elastic bodies. Romanian Academy Press, Bucharest (1972)
Teodorescu, P.P.: Mechanical Systems, Classical Models. Springer, Berlin (2006)

Chapter 6
Modal Analysis

6.1 Introduction

In many problems of mechanics of discrete systems, by linearizing the forces that are manifested between different components of the system, the motion equations are obtained in the form:

$$[M]\{\ddot{\Delta}\} + [K]\{\Delta\} = \{F\}, \tag{6.1}$$

or, if damping is also considered, in a more complex form:

$$[M]\{\ddot{\Delta}\} + [C]\{\dot{\Delta}\} + [K]\{\Delta\} = \{F\}. \tag{6.2}$$

In these equations, the matrix coefficients involved are considered to be constant. For some applications, these coefficients may depend on time, directly or indirectly, through the parameters determined by the system configuration, depending on time. Consequently, a study of the type of equations given in (6.1) and (6.2) can provide the solution for a large class of problems. In the following, an analysis of systems of this form will be made. The fundamental problems involved in the modal analysis are presented, i.e. a chapter that has a particular development in vibration analysis (see Tofan and Vlase 1985; Rades 1979, 2010; Bathe and Wilson 1976; Börm and Mehl 2012; Heylen et al. 1997).

6.2 Modal Analysis

The modal analysis (modal superposition method) is a powerful tool for a qualitative and quantitative analysis of differential equation systems with constant coefficients, a type of equation that is frequently encountered in engineering practice. It also provides excellent intuitive support for the physical understanding of phenomena.

S. Vlase et al., *Eigenvalue and Eigenvector Problems in Applied Mechanics*, Advanced Structured Materials 96, https://doi.org/10.1007/978-3-030-00991-5_6

This method consists of bringing the written system to a simpler form with the following linear transformation:

$$\{\Delta\} = [\Phi]\{q\}, \tag{6.3}$$

where $[\Phi]$ is called the modal matrix. In this way the system of n second-order differential equations describing the linear vibrations of the system *breaks* into a number of n second-order differential equations with constant coefficient, containing only one independent function, whose solving is no longer a problem. For Eq. (6.1), this transformation always exists, whereas for (6.2) the damping matrix has to respect certain conditions that will be mentioned. But often, in a first approach, the damping matrix can be neglected and the results obtained are very close to reality.

6.2.1 Eigenvalues—Natural Frequencies

Consider the system of Eq. (6.1). The notion of eigenfrequency (eigenvalue) for a physical system whose motion is described by these equations is introduced for the case where the system is not excited (see Bratu 2000; DenHartog 1960; Meirovitch 1986, 1996; Schmitz and Scott 2012; Timoshenko and Zoung 1955; Timishenko and Gere 2009).

According to the theory of differential equations with constant coefficients, we find solutions of the form:

$$\{\Delta(t)\} = \{X\}cos(\omega t + \psi). \tag{6.4}$$

The following relations are obtained by differentiation:

$$\{\dot{\Delta}(t)\} = -\omega\{X\}\sin(\omega t + \psi), \tag{6.5}$$

$$\{\ddot{\Delta}(t)\} = -\omega^2\{X\}\cos(\omega t + \psi). \tag{6.6}$$

Imposing to the solution to verify the system, we obtain:

$$\big([K] - \omega^2[M]\big)\{X\}\sin(\omega t + \psi) = 0, \tag{6.7}$$

at any time t. This requires:

$$\big([\mathrm{K}] - \omega^2[M]\big)\{X\} = 0, \tag{6.8}$$

which represents a linear, homogeneous system. It will admit other solutions than the trivial solution (zero solution) if and only if:

$$\det\big([K] - \omega^2[M]\big) = 0. \tag{6.9}$$

If n is the size of the system (number of differential equations), the evaluation of the determinant (6.9) results in a n-degree polynomial in ω^2:

$$P(\omega^2) = \det\big([K] - \omega^2[M]\big). \tag{6.10}$$

This is called the characteristic polynomial of the system. The equation $P(\lambda) = 0$ is called the characteristic equation or the frequency equation or the eigenvalue equation. The quantities $\omega_i = \sqrt{\lambda_i}$ are called the system's natural frequencies (circular eigenfrequencies). The definitions for circular frequencies and periods are the same as for a single-degree of freedom vibration studies. The lowest eigenfrequency of the system is called the fundamental frequency and usually plays a fundamental role in applications.

Eigenvalues/eigenfrequencies/eigenperiods can be ordered, defining spectral series consisting of:

- eigenvalues

$$\{\omega\} = [\omega_1\ \omega_2\ \ldots\ \omega_n]^T,$$

  with ω_i ordered ascending: $\omega_1 < \omega_2 < \ldots < \omega_n$;
- eigenfrequencies $f_i = \frac{\omega_i}{2\pi}$
  with: $f_1 < f_2 < \ldots < f_n$;
- eigenperiods $T_i = \frac{2\pi}{\omega_i}$
  with: $T_1 > T_2 > \ldots > T_n$.

The spectrum of eigenvalues/eigenfrequencies/eigenperiods defines the natural characteristics of the system and depends only on the distribution of masses and the elastic properties of the elements that compose the studied system. It is not influenced in any way by the initial values of positions and speeds, nor by forces acting on the system. The number of the eigenvalues is equal to the number of degrees of freedom of the system.

We note the solution of the linear homogeneous system (6.8), in which we introduced the eigenvalues ω_i previously determined, with $\{X_i\}$:

$$\big([K] - \omega_i^2[M]\big)\{X_i\} = 0 \text{ for } \quad i = \overline{1, n}. \tag{6.11}$$

This solution is called the eigenvector of rank i of the system. Whereas:

$$\det\big([K] - \omega_i^2[M]\big) = 0,$$

the n components of the vector $\{X_i\}$ will not be independent, so we will not be able to get the modulus of the vector but only its direction. It is possible to obtain the *mode of motion* of the rank i if a way of scaling (normalization) of the components $\{X_i\}$ is established. It is denoted, after scaling, the eigenmodes with $\{\Phi_i\} = \mu_i\{X_i\}$ and their ensemble form the modal matrix $[\Phi]$:

$$[\Phi] = [\Phi_1\ \Phi_2 \ldots \Phi_n]. \tag{6.12}$$

The scalar depends on how the normalization of the components of $\{X_{\rm i}\}$ is performed. Some procedures will be presented in the next paragraph.

6.2.2 *Properties of the Eigenvalues*

P1 If the matrix $[K]$ is singular, then $\omega^2 = 0$ is the eigenvalue of the characteristic equation.

Proof Indeed $P(0) = \det([K] - 0 \cdot [M]) = \det[K] = 0$, so $\omega^2 = 0$ is therefore the eigenvalue.

Furthermore, if the rank$(K) = n - l$, then $^2 = 0$ is the root of the equation $P(\omega^2) = 0$ with the order l of multiplicity.

P2 If $[K]$ and $[M]$ are symmetrical and positive defined, then their eigenvalues are real.

Proof If the equation with real coefficients det $([K] - \lambda\ [M]) = 0$ admits λ as a root, then $\bar{\lambda}$ admits as well. We have successively:

$$[K]\{\Phi_i\} = \lambda_i[M]\{\Phi_i\}, \tag{6.13}$$

$$[K]\{\overline{\Phi}_i\} = \bar{\lambda}_i[M]\{\overline{\Phi}_i\}. \tag{6.14}$$

If we premultiply (6.13) with $\{\overline{\Phi}_i\}^T$ and (6.14) with $\{\Phi_i\}^T$, one obtains, respectively:

$$\{\overline{\Phi}_i\}^T[K]\{\Phi_i\} = \lambda_i\{\overline{\Phi}_i\}^T[M]\{\Phi_i\}, \tag{6.15}$$

$$\{\Phi_i\}^T[K]\{\overline{\Phi}_i\} = \bar{\lambda}_i\{\Phi_i\}^T[M]\{\overline{\Phi}_i\}. \tag{6.16}$$

By virtue of the symmetry of the mass and rigidity matrix, it will be obtained by substructing:

$$0 = (\lambda_i - \bar{\lambda}_i)\{\Phi_i\}^T[M]\{\overline{\Phi}_i\}, \tag{6.17}$$

from where, generally $\lambda_i = \bar{\lambda}_i$, because $\{\Phi_i\}^T[M]\{\overline{\Phi}_i\}$ it is a definite positive quadratic form. It turns out that λ_i will be real.

We give, without demonstration, the following property.

P3 If $[K]$ and $[M]$ are positively defined, then their eigenvalues are positive. For mechanical systems studied in our presentation this always happens.

6.2.3 Orthogonality Properties

Since the matrix $[K]$ is symmetric we can write:

$$\{X_i\}^T[K]\{X_j\} = \{X_j\}^T[K]\{X_i\}, \tag{6.18}$$

a relationship that is readily demonstrated by the explicit writing of matrices that appear in written formula.

Since the matrix $[M]$ is symmetric, we also have:

$$\{X_i\}^T[K]\{X_j\} = \{X_j\}^T[M]\{X_i\}. \tag{6.19}$$

We write the equation of eigenvalues as:

$$[K]\{X_i\} - \omega_i^2[M]\{X_i\} = 0. \tag{6.20}$$

If we premultiply with $\{X_j\}^T$, one gets:

$$\{X_j\}^T[K]\{X_i\} - \omega_i^2\{X_j\}^T[M]\{X_i\} = 0. \tag{6.21}$$

Analogously we can also obtain:

$$\{X_i\}^T[K]\{X_j\} - \omega_j^2\{X_i\}^T[M]\{X_j\} = 0. \tag{6.22}$$

By subtracting relations (6.21) and (6.22) we obtain:

$$(\omega_i^2 - \omega_j^2)\{X_j\}^T[M]\{X_i\} = 0. \tag{6.23}$$

Because, generally, $\omega_i \neq \omega_j$ it results:

$$\{X_j\}^T[M]\{X_i\} = 0, \tag{6.24}$$

so too:

$$\{X_j\}^T[K]\{X_i\} = 0. \tag{6.25}$$

We can take instead of the amplitudes $\{X\}$ the eigenmodes and get:

$$\{\Phi_j\}^T[M]\{\Phi_i\} = 0$$

and

$$\{\Phi_j\}^T[K]\{\Phi_i\} = 0.$$

We say that the eigenmodes are orthogonal through $[M]$, respectively $[K]$.

6.2.4 Rayleigh's Quotient

If in the relationship:

$$[K]\{X_i\} - \omega_i^2[M]\{X_i\} = 0, \tag{6.26}$$

we premultiply to the left with $\{X_i\}^T$, we get:

$$\{X_i\}^T[K]\{X_i\} - \omega_i^2\{X_i\}^T[M]\{X_i\} = 0, \tag{6.27}$$

from where:

$$\omega_i^2 = \frac{\{X_i\}^T[K]\{X_i\}}{\{X_i\}^T[M]\{X_i\}}. \tag{6.28}$$

The written relationship is called the Rayleigh's quotient. It can be shown that, if the approximate (estimated) eigenvectors $\{X_i\}$ are chosen, a very good approximation can be obtained for ω_i^2. It is the reason that this ratio is used for the approximate calculation of eigenfrequencies. In the above relationship $\{X_i\}$ can be replaced by $\{\Phi_i\}$.

6.2.5 Generalized Orthogonality Relationships

From the relationship

$$[K]\{\Phi_j\} - \omega_j^2[M][\Phi_j] = 0, \tag{6.29}$$

one can express $\{\Phi_j\}$ in two distinct ways:

(i)

$$\{\Phi_j\} = \frac{1}{\omega_j^2}[M]^{-1}[K]\{\Phi_j\}; \tag{6.30}$$

(ii)

$$\{\Phi_j\} = \omega_j^2[K]^{-1}[M]\{\Phi_j\} = \omega_j^2([M]^{-1}[K])^{-1}\{\Phi_j\}. \tag{6.31}$$

Taking into account these two relationships, the orthogonality properties can be generalized in two complementary directions.

Thus, if the relationship shown above is considered:

$$\{\Phi_i\}^T[M]\{\Phi_j\} = 0 \tag{6.32}$$

and $\{\Phi_j\}$ is introduced with the relation (i), it can be written:

$$\{\Phi_i\}^T[M]\frac{1}{\omega_j^2}[M]^{-1}[K]\{\Phi_j\} = 0,$$

or

$$\{\Phi_i\}^T[M]\big([M]^{-1}[K]\big)\{\Phi_j\} = 0.$$

Replacing $\{\Phi_j\}$ from (i) one obtains:

$$\{\Phi_i\}^T[M]\big([M]^{-1}[K]\big)\frac{1}{\omega_j^2}[M]^{-1}[K]\{\Phi_j\} = 0,$$

or

$$\{\Phi_i\}^T[M]\big([M]^{-1}[K]\big)^2\{\Phi_j\} = 0.$$

Repeating the process by a mathematical induction process:

$$\{\Phi_i\}^T[M]\big([M]^{-1}[K]\big)^p\{\Phi_j\} = 0, \tag{6.33}$$

where the relationship is valid for $p = 0, 1, 2, 3, \ldots$

Further, if in (6.32) we introduce $\{\Phi_j\}$ expressed with relation (ii), we can write:

$$\{\Phi_i\}[M]\omega_j^2[K]^{-1}[M]\{\Phi_j\} = 0,$$

or

$$\{\Phi_i\}[M]\big([M]^{-1}[K]\big)^{-1}\{\Phi_j\} = 0.$$

If in this relationship $\{\Phi_j\}$ is reintroduced:

$$\{\Phi_i\}[M]\big([M]^{-1}[K]\big)^{-1}\omega_j^2[K]^{-1}[M]\{\Phi_j\} = 0,$$

or

$$\{\Phi_i\}[M]\big([M]^{-1}[K]\big)^{-2}\{\Phi_j\} = 0.$$

Repeating this process using the mathematical induction we obtain:

$$\{\Phi_i\}^T[M]\big([M]^{-1}[K]\big)^{-p}\{\Phi_j\} = 0, \tag{6.34}$$

where the relationship is valid for $p = 0, 1, 2, 3, \ldots$

The two relationships (6.33) and (6.34) can be grouped into one, which represents the generalized orthogonality properties:

$$\{\Phi_i\}^T[M]\big([M]^{-1}[K]\big)^{p}\{\Phi_j\} = 0, \quad p \in Z. \tag{6.35}$$

6.2.6 *Definition of Relationships for the Damping Matrix*

The Caughey damping matrix is defined as a linear combination of orthogonal matrices $[M]\big([M]^{-1}[K]\big)^{p}$:

$$[C] = \sum_{r \in Z} \delta_r [M]\big([M]^{-1}[K]\big)^{r} \quad \text{with} : \alpha_r \in R. \tag{6.36}$$

For a system with n degrees of freedom this linear combination contains only n independent terms. Thus, it can be written:

$$[C] = \sum_{r=0}^{n-1} \alpha_r [M]\big([M]^{-1}[K]\big)^{r} = [M]\sum_{r=0}^{n-1} \alpha_r \big([M]^{-1}[K]\big)^{r}. \tag{6.37}$$

The Rayleigh damping matrix is obtained as a particular case of the Caughey matrix by retaining the first two terms:

$$[C] = [C_0] + [C_2] = \alpha_0[M] + \alpha_1[K]. \tag{6.38}$$

6.2.7 *Normalized Vibration Modes*

The solution of the eigenvalue problem consists of the uniquely determined vector of eigenvalues and the vector system of the amplitudes, determined up to a factor. The process by which the undetermination introduced by the scale factor is eliminated is called the normalization. The normalization can be done in several ways. Some of the used methods are outlined in the following:

- consider the amplitude of the first degree of freedom equal to unity;
- consider the maximum amplitude of each vibration mode equal to unity;

- impose the condition $\{X_i\}^T[M]\{X_i\} = 1$ that defines the form normalized by $[M]$ of the vibration mode i. This type of normalization is called the inertial normalization;
- impose the condition $\{X_i\}^T\{X_i\} = 1$, so that the sum of the squares of the amplitudes is 1 (the eigenvectors are unit vectors).

Normalizing the eigenvectors (eigendirections) means that they become uniquely determined (eigenmodes of vibration).

6.2.8 Decoupling the Motion Equations

The system of differential equations describing the motion of a mechanical system with concentrated parameters is:

$$[M]\{\ddot{\Delta}\} + [C]\{\dot{\Delta}\} + [K]\{\Delta\} = \{F\}, \tag{6.39}$$

which becomes, with the transformation $\{\Delta\} = [\Phi]\{q\}$, a system of differential equations in $\{q\}$:

$$[M][\Phi]\{\ddot{q}\} + [C][\Phi]\{\dot{q}\} + [K][\Phi]\{q\} = \{F\}. \tag{6.40}$$

If the damping matrix $[C]$ is a matrix that admits a Caughey factorization, then the system of equations will decouple by multiplying them with the transposed of the modal matrix:

$$[\Phi]^T[M][\Phi]\{\ddot{q}\} + [\Phi]^T[C][\Phi]\{\dot{q}\} + [\Phi]^T[K][\Phi]\{q\} = [\Phi]^T\{F\}. \tag{6.41}$$

The matrices $[\Phi]^T[M][\Phi]$, $[\Phi]^T[C][\Phi]$ and $[\Phi]^T[K][\Phi]$ become diagonal in this case, and we denote them as:

$$\begin{aligned}
[\backslash M^*\backslash] &= [\Phi]^T[M][\Phi],\\
[\backslash C^*\backslash] &= [\Phi]^T[C][\Phi],\\
[\backslash K^*\backslash] &= [\Phi]^T[K][\Phi].
\end{aligned} \tag{6.42}$$

It is noted with m_i, c_i and respectively k_i, the diagonal entries of the matrices $[\backslash M^*\backslash]$, $[\backslash C^*\backslash]$ and $[\backslash K^*\backslash]$. These are obtained using the relationships:

$$\begin{aligned}
m_i &= [\Phi_i]^T[M][\Phi_i],\\
c_i &= [\Phi_i]^T[C][\Phi_i],\\
k_i &= [\Phi_i]^T[K][\Phi_i].
\end{aligned} \tag{6.43}$$

With these notations, the system of differential equations is decoupled into n second-order differential equations with constant coefficients, containing only one unknown functions q_i:

$$\begin{aligned} m_1\ddot{q}_1 + c_1\dot{q}_1 + k_1 q_1 &= F_1, \\ m_2\ddot{q}_2 + c_2\dot{q}_2 + k_2 q_2 &= F_2, \\ \ldots \\ m_n\ddot{q}_n + c_n\dot{q}_n + k_n q_n &= F_n. \end{aligned} \tag{6.44}$$

With the notations: $2\beta_i = c_i/m_i$, $\omega_i^2 = k_i/m_i$, and $f_i = F_i/m_i$ there is obtained:

$$\begin{aligned} \ddot{q}_1 + 2\beta_1\dot{q}_1 + \omega_1^2 q_1 &= f_1, \\ \ddot{q}_2 + 2\beta_2\dot{q}_2 + \omega_2^2 q_2 &= f_2, \\ \ldots \\ \ddot{q}_n + 2\beta_n\dot{q}_n + \omega_n^2 q_n &= f_n. \end{aligned} \tag{6.45}$$

The problem is thus reduced to solving n independent second order differential equations. The first step to solve this system is by writing the homogeneous equation:

$$\ddot{q}_i + 2\beta_i\dot{q}_i + \omega_i^2 q_i = 0_i, \tag{6.46}$$

which has the homogeneous solution:

$$q_{io} = A_i \exp(-i\beta t)\sin(\omega_i t + \psi_i), \tag{6.47}$$

to which a particular q_{ip} solution will be added:

$$q_i = q_{io} + q_{ip}.$$

The solution of the differential system is obtained by modal superposition:

$$\{\Delta_o\} = \{\Phi_1\}q_1 + \{\Phi_2\}q_2 + \ldots + \{\Phi_n\}q_n. \tag{6.48}$$

Another way to determine a particular solution, if the force vector admits a harmonic decomposition, can be done directly on the initial system. If we consider the system:

$$[M]\{\ddot{\Delta}\} + [C]\{\dot{\Delta}\} + [K]\{\Delta\} = \{F_{ic}\}\cos ipt + F_{is}\}\sin ipt, \tag{6.49}$$

in which, for the force vector we have considered only one of the harmonics, p, a particular solution can be obtained in the form of:

$$\{\Delta_{ip}\} = \{X_{ic}\}\cos ipt + \{X_{is}\}\sin ipt. \tag{6.50}$$

By differentiation, one obtains:

$$\{\dot{\Delta}_{ip}\} = -p\{X_{ic}\}\sin ipt + p\{X_{is}\}\cos ipt \tag{6.51}$$

and

$$\{\ddot{\Delta}_{ip}\} = -p^2\{X_{ic}\}\cos ipt + p^2\{X_{is}\}\sin ipt. \tag{6.52}$$

If we replace these relationships in the previously written differential equation system (6.49), we obtain:

$$\begin{aligned}
&[M]\left(-i^2p^2\{X_{ic}\}\cos ipt - i^2p^2\{X_{is}\}\sin ipt\right) \\
&\quad + [C](-ip\{X_{ic}\}\sin ipt + ip\{X_{is}\}\cos ipt) \\
&\quad + [K](\{X_{ic}\}\cos ipt + \{X_{is}\}\sin ipt) \\
&= \{F_{ic}\}\cos ipt + \{F_{ic}\}\sin ipt,
\end{aligned}$$

or

$$\begin{aligned}
&\left(-i^2p^2[M]\{X_{ic}\} + ip[C]\{X_{ic}\} + [K]\{X_{ic}\} - \{F_{ic}\}\right)\cos ipt \\
&\quad + \left(-i^2p^2[M]\{X_{is}\} - ip[C]\{X_{1c}\} + [K]\{X_{is}\} - \{F_{ic}\}\right)\sin ipt = 0.
\end{aligned} \tag{6.53}$$

The written relations must be valid at any time t, it results that the coefficients of the terms cos pt and sin pt must be zero, so we must have:

$$\begin{cases}
\left([K] - i^2p^2[M]\right)\{X_{1c}\} + ip[C]\{X_{is}\} = \{F_{ic}\} \\
-p[C]\{X_{ic}\}\left([K] - i^2p^2[M]\right)\{X_{is}\} = \{F_{ic}\}
\end{cases},$$

or

$$\begin{bmatrix} [K] - i^2p^2[M] & ip[C] \\ -ip[C] & [K] - i^2p^2[M] \end{bmatrix} \begin{Bmatrix} \{X_{ic}\} \\ \{X_{is}\} \end{Bmatrix} = \begin{Bmatrix} \{F_{ic}\} \\ \{F_{is}\} \end{Bmatrix}. \tag{6.54}$$

The system solution is obtained as:

$$\begin{Bmatrix} \{X_{ic}\} \\ \{X_{is}\} \end{Bmatrix} \begin{bmatrix} [K] - i^2p^2[M] & ip[C] \\ -ip[C] & [K] - i^2p^2[M] \end{bmatrix}^{-1} = \begin{Bmatrix} \{F_{ic}\} \\ \{F_{is}\} \end{Bmatrix}. \tag{6.55}$$

Analogous solutions for the other harmonics of the Fourier decomposition of force can be obtained, the particular solution being the overlapping of all these solutions:

$$\{\Delta_{par}\} = \{\Delta_{st}\} + \{\Delta_{1p}\} + \{\Delta_{2p}\} + \ldots + \{\Delta_{np}\}. \tag{6.56}$$

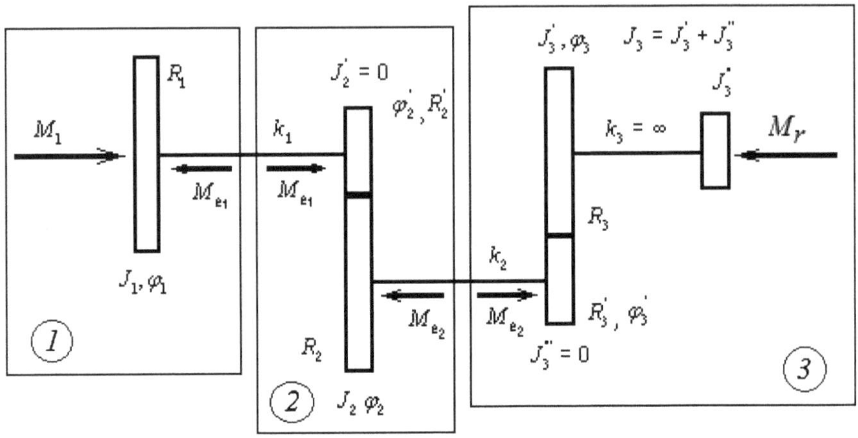

Fig. 6.1 Elastic system with three degrees of freedom

In most engineering applications, the solution of the homogeneous damped equation disappears after a reasonable period of time and what matters is the particular solution obtained with (6.56).

6.3 Application

A1 Consider a transmission (such as a gearbox) with two gear trains, presented in Fig. 6.1. The task is to study the torsional vibration of this transmission.

The torque produced by an engine, M_1, is transmitted to the flywheel with the moment of inertia J_1 and radius R_1, from where, through the elastic shaft with the rigidity k_1, it is transmitted to the toothed wheel with the moment of inertia $J_2' = 0$ and the radius R_2'. The torque is then transmitted through the gear to the toothed wheel with the moment of inertia J_2 and the radius R_2. Then using the elastic shaft with the rigidity k_2, the torque is transmitted to the first toothed wheel with the radius R_3' and a negligible moment of inertia $J_3' = 0$.

This torque is transmitted through the gear to the toothed wheel with the moment of inertia J_3 and radius R_3. Then, through a rigid shaft ($k_3 = \infty$), to the wheel with radius R_4 and moment of inertia J_4 where a resistant torque acts, M_{r}. The transmission can be reduced to a system with three degrees of freedom.

(i) *Equations of motion*

In Fig. 6.1 an elastic system is presented and its division into subsystems. The following notations will be made:

$$i_2 = \frac{R_2}{R_2'}, \quad i_3 = \frac{R_3}{R_3'} \times \frac{R_2}{R_2'}, \quad \left(i_3 = i_2 \frac{R_3'}{R_3}\right),$$

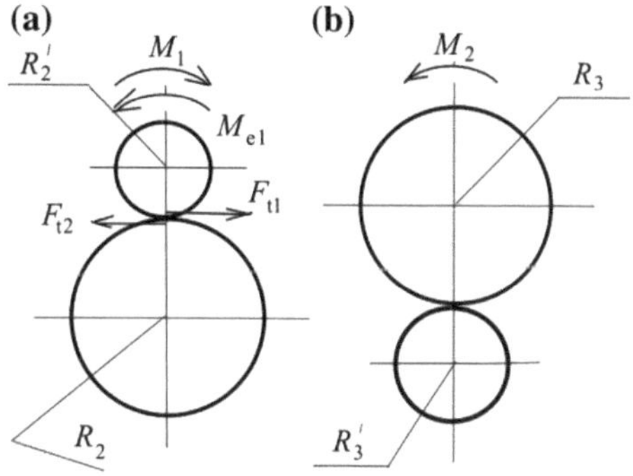

Fig. 6.2 Free body diagram

where i_2 and i_2 are the ratio of the rotational speed of the shafts determined by the two gears. Then, using the d'Alembert equations (see Wittenburg 2008; Vlase 1987a, b, 2003; Negrean 2017a, b), the motion equation for the wheel of the subsystem 1 is:

$$M_1 - M_{e1} - M_1^i = 0, \tag{6.57}$$

where M_1 is the torque of an engine that drives the transmission, $M_{e1} = k_1(\varphi_1 - \varphi_2') = k_1(\varphi_1 - i_2\varphi_2)$ is the elastic moment that occurs in the shaft between the flywheel 1 and the flywheels 2′ and $M_1^i = J_1\ddot{\varphi}_1$ is the torque due to the inertia of the flywheel *1*. In this case the Eq. (6.57) can be written:

$$J_1\ddot{\varphi}_1 + k_1(\varphi_1 - i_2\varphi_2) = M_1. \tag{6.57'}$$

The angle φ_2' of the toothed wheel with the radius R_2' becomes, due to the gear, the angle $\varphi_2 = \frac{R_2'}{R_2}\varphi_2' = \frac{1}{i_2}\varphi_2'$ of the toothed wheel with the radius R_2'. We considered that the moment of inertia of the flywheel with the radius R_2' is zero, so the inertia of this flywheel has no influence in the motion equations, having only a kinematic significance.

The motion equation for the flywheel with the radius R_2 (Fig. 6.2a—system 2) is:

$$i_2 M_{e1} - M_{e2} - M_2^i = 0$$

or

$$J_2\ddot{\varphi}_2 - i_2(\varphi_1 - i_2\varphi_2)k_1 + k_2\left(\varphi_2 - \frac{i_3}{i_2}\varphi_3\right) = 0.$$

The moment M_{e1} has become (by virtue of the fact that the tangential force in the gear is preserved—$F_{t1} = F_{t2}$), iM_{e1}. From:

$$F_{t1} = F_{t2},$$

it results:

$$\frac{M_{e1}}{R_2'} = \frac{M_{e1}'}{R_2},$$

from where:

$$M_{e1}' = \frac{R_2}{R_2'} M_{e1} = i M_{e1}.$$

For the third wheel (the system 3) it can be written:

$$\frac{i_3}{i_2} M_{e2} - M_2 - M_3^i = 0,$$

or

$$J_3 \ddot{\varphi}_3 - \frac{i_3}{i_2}\left(\varphi_2 - \frac{i_3}{i_2}\varphi_3\right) k_2 = -M_2.$$

The three equations form the system of differential equations are:

$$\begin{aligned} &J_1 \ddot{\varphi}_1 + k_1(\varphi_1 - i_2\varphi_2) = M_1, \\ &J_2 \ddot{\varphi}_2 - i_2 k_1(\varphi_1 - i_2\varphi_2) + k_2\left(\varphi_2 - \frac{i_3}{i_2}\varphi_3\right) = 0, \\ &J_3 \ddot{\varphi}_3 - k_2 \frac{i_3}{i_2}\left(\varphi_2 - \frac{i_3}{i_2}\varphi_3\right) = -M_2, \end{aligned}$$

or, in matrix form:

$$\begin{bmatrix} J_1 & 0 & 0 \\ 0 & J_2 & 0 \\ 0 & 0 & J_3 \end{bmatrix} \begin{Bmatrix} \ddot{\varphi}_1 \\ \ddot{\varphi}_2 \\ \ddot{\varphi}_3 \end{Bmatrix} + \begin{bmatrix} k_1 & -i_2 k_1 & 0 \\ -i_2 k_1 & i_2^2 k_1 + k_2 & -\frac{i_3}{i_2} k_2 \\ 0 & -\frac{i_3}{i_2} k_2 & \frac{i_3^2}{i_2^2} k_2 \end{bmatrix} \begin{Bmatrix} \varphi_1 \\ \varphi_2 \\ \varphi_3 \end{Bmatrix} = \begin{Bmatrix} M_1 \\ 0 \\ -M_2 \end{Bmatrix}.$$

(ii) *Equivalent model*

Based on Fig. 6.1 and the previous analysis for the given elastic system, we can draw the figure of the equivalent system (see Fig. 6.3), consisting of three elastically coupled flywheels.

(iii) *Changing coordinates*

If the system behaves as a rigid body, the rigid motion mode is:

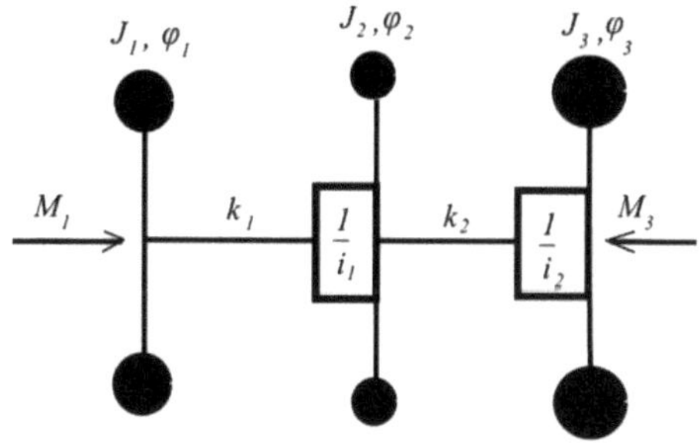

Fig. 6.3 System with three wheels

$$\begin{Bmatrix} \varphi_1 \\ \varphi_2 \\ \varphi_3 \end{Bmatrix} = \begin{Bmatrix} 1 \\ \frac{1}{i_2} \\ \frac{1}{i_3} \end{Bmatrix} \varphi_1 .$$

It is natural to refer the motion of all the wheels to the first wheel drive through the new functions:

$$\begin{Bmatrix} \theta_1 \\ \theta_2 \\ \theta_3 \end{Bmatrix} = \begin{bmatrix} 1 & 0 & 0 \\ 0 & i_2 & 0 \\ 0 & 0 & i_3 \end{bmatrix} \begin{Bmatrix} \varphi_1 \\ \varphi_2 \\ \varphi_3 \end{Bmatrix},$$

from where it results:

$$\begin{Bmatrix} \varphi_1 \\ \varphi_2 \\ \varphi_3 \end{Bmatrix} = \begin{bmatrix} 1 & 0 & 0 \\ 0 & \frac{1}{i_2} & 0 \\ 0 & 0 & \frac{1}{i_3} \end{bmatrix} \begin{Bmatrix} \theta_1 \\ \theta_2 \\ \theta_3 \end{Bmatrix}.$$

In this case the motion equations become:

$$\begin{bmatrix} J_1 & 0 & 0 \\ 0 & J_2 & 0 \\ 0 & 0 & J_3 \end{bmatrix} \begin{bmatrix} 1 & 0 & 0 \\ 0 & \frac{1}{i_2} & 0 \\ 0 & 0 & \frac{1}{i_3} \end{bmatrix} \begin{Bmatrix} \ddot{\theta}_1 \\ \ddot{\theta}_2 \\ \ddot{\theta}_3 \end{Bmatrix}$$

$$+ \begin{bmatrix} k_1 & -i_2 k_1 & 0 \\ -i_2 k_1 & i_2^2 k_1 + k_2 & -\frac{i_3}{i_2} k_2 \\ 0 & -\frac{i_3}{i_2} k_2 & \frac{i_3^2}{i_2^2} k_2 \end{bmatrix} \begin{bmatrix} 1 & 0 & 0 \\ 0 & \frac{1}{i_2} & 0 \\ 0 & 0 & \frac{1}{i_3} \end{bmatrix} \begin{Bmatrix} \theta_1 \\ \theta_2 \\ \theta_3 \end{Bmatrix} = \begin{Bmatrix} M_1 \\ 0 \\ -M_2 \end{Bmatrix}.$$

In order to transform the forces and moments into the space determined by the new independent coordinates $[\,\theta_1 \; \theta_2 \; \theta_3\,]$, the equations are premultiplied with:

Fig. 6.4 Equivalent mathematical model

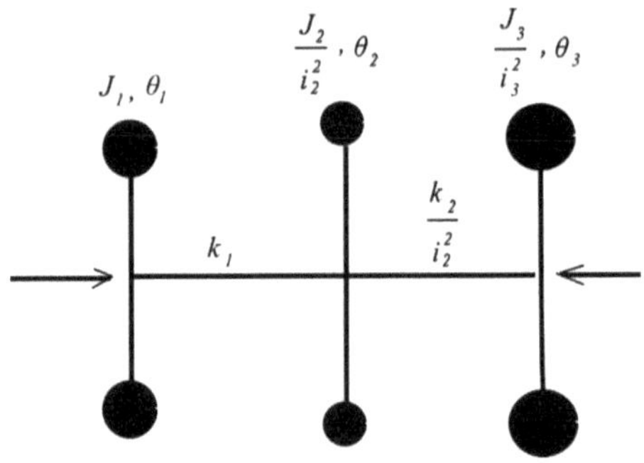

$$\begin{bmatrix} 1 & 0 & 0 \\ 0 & \frac{1}{i_2} & 0 \\ 0 & 0 & \frac{1}{i_3} \end{bmatrix}^T.$$

The following equations are obtained:

$$\begin{bmatrix} J_1 & 0 & 0 \\ 0 & \frac{J_2}{i_2^2} & 0 \\ 0 & 0 & \frac{J_3}{i_3^2} \end{bmatrix} \begin{Bmatrix} \ddot{\theta}_1 \\ \ddot{\theta}_2 \\ \ddot{\theta}_3 \end{Bmatrix} + \begin{bmatrix} k_1 & -k_1 & 0 \\ -k_1 & k_1 + \frac{k_2}{i_3^2} & -\frac{k_2}{i_2^2} \\ 0 & -\frac{k_2}{i_2^2} & \frac{k_2}{i_2^2} \end{bmatrix} \begin{Bmatrix} \theta_1 \\ \theta_2 \\ \theta_3 \end{Bmatrix} = \begin{Bmatrix} M_1 \\ 0 \\ -\frac{M_2}{i_3} \end{Bmatrix}.$$

The equivalent mathematical model is shown in Fig. 6.4.

(iv) *Numerical simulation*

For a numerical calculation, we choose the following system of values for the elements that define the system:

$$J_1 = 3J, \frac{J_2}{i_2^2} = 2J, \frac{J_3}{i_3^2} = 2J, k_1 = 3k, \frac{k_2}{i_2^2} = 2k.$$

With these values the motion equations take the form:

$$J\begin{bmatrix} 3 & 0 & 0 \\ 0 & 2 & 0 \\ 0 & 0 & 1 \end{bmatrix} \begin{Bmatrix} \ddot{\theta}_1 \\ \ddot{\theta}_2 \\ \ddot{\theta}_3 \end{Bmatrix} + k\begin{bmatrix} 3 & -3 & 0 \\ -3 & 5 & -2 \\ 0 & -2 & 2 \end{bmatrix} \begin{Bmatrix} \theta_1 \\ \theta_2 \\ \theta_3 \end{Bmatrix} = \begin{Bmatrix} M_1 \\ 0 \\ -\frac{M_2}{i_3} \end{Bmatrix}.$$

The dynamic matrix is:

$$[p^2] = [J]^{-1}[K] = \frac{k}{2J}\begin{bmatrix} 2 & -2 & 0 \\ -3 & 5 & -2 \\ 0 & -2 & 2 \end{bmatrix}.$$

(v) *Eigenvalues*

The characteristic equation is:

$$\det\left([p^2] - \omega^2[E]\right) = 0 \quad \left(\text{from } \det\left([K] - \omega^2[J]\right) = 0\right).$$

It results:

$$\begin{vmatrix} 2-\lambda & -2 & 0 \\ -3 & 5-\lambda & -2 \\ 0 & -2 & 2-\lambda \end{vmatrix} = 0,$$

where the abbreviation $\lambda = \omega^2 \frac{2J}{k}$ was used. Thus, the following equations follows:

$$\lambda^3 - 9\lambda^2 + 14\lambda = 0,$$

with the solutions:

$$\lambda_1 = 0, \quad \lambda_2 = 2, \quad \lambda_3 = 7.$$

From here comes the circular eigenfrequencies:

$$\begin{Bmatrix} \omega_1 \\ \omega_2 \\ \omega_3 \end{Bmatrix} = \sqrt{\frac{k}{2J}} \begin{Bmatrix} 0 \\ \sqrt{2} \\ \sqrt{7} \end{Bmatrix}.$$

(vi) *Eigenmodes*

For the first eigenvalue $\omega_1^2 = 0$, from the homogeneous linear system the rigid motion mode will result:

$$\begin{vmatrix} 2 & -2 & 0 \\ -3 & 5 & -2 \\ 0 & -2 & 2 \end{vmatrix} \begin{Bmatrix} e_1 \\ e_2 \\ e_3 \end{Bmatrix} = \{0\}.$$

If one chooses $e_1 = 1$ we get for the first mode of motion:

$$\{\Phi_1\} = \begin{Bmatrix} 1 \\ 1 \\ 1 \end{Bmatrix}.$$

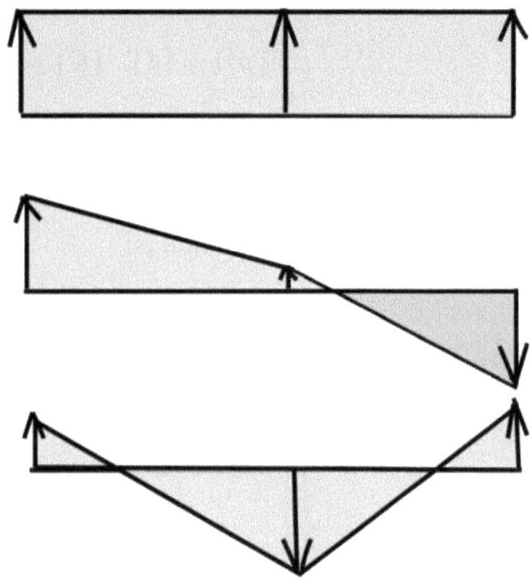
Fig. 6.5 Eigenmodes

For the second eigenvalue $\omega_2^2 = \frac{k}{J}$, it results the system:

$$\begin{vmatrix} 0 & -2 & 0 \\ -3 & 3 & -2 \\ 0 & -2 & 0 \end{vmatrix} \begin{Bmatrix} e_1 \\ e_2 \\ e_3 \end{Bmatrix} = \{0\}.$$

If we choose $e_1 = 1$, then the second eigenmode is:

$$\{\Phi_2\} = \begin{Bmatrix} 1 \\ 0 \\ -1.5 \end{Bmatrix}.$$

Finally, for the third eigenvalue $\omega_2^2 = \frac{7k}{2J}$, it results the linear system:

$$\begin{vmatrix} -5 & -2 & 0 \\ -3 & -2 & -2 \\ 0 & -2 & -5 \end{vmatrix} \begin{Bmatrix} e_1 \\ e_2 \\ e_3 \end{Bmatrix} = \{0\},$$

from which the third eigenmode it results (Fig. 6.5):

$$\{\Phi_3\} = \begin{Bmatrix} 1 \\ -2.5 \\ 1 \end{Bmatrix}.$$

The modal matrix will be:

$$[\Phi] = \begin{bmatrix} 1 & 1 & 1 \\ 1 & 0 & -2.5 \\ 1 & -1.5 & 1 \end{bmatrix}.$$

(vii) *Canonical coordinate*

Let us make the transition to canonical coordinates using the transformation:

$$\begin{Bmatrix} \theta_1 \\ \theta_2 \\ \theta_3 \end{Bmatrix} = [\Phi] \begin{Bmatrix} q_1 \\ q_2 \\ q_3 \end{Bmatrix} = \begin{bmatrix} 1 & 1 & 1 \\ 1 & 0 & -2.5 \\ 1 & -1.5 & 1 \end{bmatrix} \begin{Bmatrix} q_1 \\ q_2 \\ q_3 \end{Bmatrix}.$$

The matrix coefficients of the system of equations in canonical coordinates are:

$$[\backslash M \backslash] = [\Phi]^T [M][\Phi] = J \begin{bmatrix} 7 & & 0 \\ & 7.5 & \\ 0 & & 70 \end{bmatrix},$$

$$[\backslash K \backslash] = [\Phi]^T [K][\Phi] = k \begin{bmatrix} 0 & & 0 \\ & 7.5 & \\ 0 & & 245 \end{bmatrix},$$

$$\{Q\} = [\Phi]^T \begin{Bmatrix} M_1 \\ 0 \\ -\frac{M_2}{i_3} \end{Bmatrix}.$$

In the new canonical coordinates the motion equations will take the form:

$$J \begin{bmatrix} 7 & & \\ & 7.5 & \\ & & 70 \end{bmatrix} \begin{Bmatrix} \ddot{q}_1 \\ \ddot{q}_2 \\ \ddot{q}_3 \end{Bmatrix} + k \begin{bmatrix} 0 & & \\ & 7.5 & \\ & & 245 \end{bmatrix} \begin{Bmatrix} q_1 \\ q_2 \\ q_3 \end{Bmatrix} = \begin{Bmatrix} M_1 - M_2' \\ M_1 + 2.5M_2' \\ M_1 + M_2' \end{Bmatrix},$$

where: $M_2' = \frac{M_2}{i_3}$. The system can be written in the alternative form:

$$\ddot{q}_1 = \frac{M_1 - M_2'}{7J}, \ddot{q}_2 + \frac{k}{J} q_2 = \frac{M_1 + 2.5M_2'}{J}, \ddot{q}_3 = +\frac{7k}{2J} q_3 - \frac{M_1 + M_2'}{70J}.$$

(viii) *Inertial normalization*

If we impose the condition that the normalized modal matrix respects the relationship:

$$[\Phi_M]^T[M][\Phi_M] = [E],$$

it result for the inertial normalized modal matrix:

$$[\Phi_M] = [\Phi][M]^{-\frac{1}{2}} = \begin{bmatrix} 1 & 1 & 1 \\ 1 & 0 & -2.5 \\ 1 & -1.5 & 1 \end{bmatrix} \begin{bmatrix} \frac{1}{\sqrt{7}} & 0 & 0 \\ 0 & \frac{1}{\sqrt{7.5}} & 0 \\ 0 & 0 & \frac{1}{\sqrt{70}} \end{bmatrix}$$

$$= \begin{bmatrix} 0.3785 & 0.365 & 0.239 \\ 0.3785 & 0 & -0.597 \\ 0.3785 & -0.548 & 0.239 \end{bmatrix}.$$

(ix) *Spectral matrix*

The equations of motion in the space of canonical coordinates, with inertial normalization, become:

$$\begin{Bmatrix} \ddot{q}_1 \\ \ddot{q}_2 \\ \ddot{q}_3 \end{Bmatrix} + \frac{k}{2J} \begin{bmatrix} 0 & 0 & 0 \\ 0 & 2 & 0 \\ 0 & 0 & 7 \end{bmatrix} \begin{Bmatrix} q_1 \\ q_2 \\ q_3 \end{Bmatrix} = \frac{1}{J} \begin{Bmatrix} 0.3785(M_1 - M_2') \\ 0.365M_1 + 0.548M_2' \\ 0.239(M_1 + M_2') \end{Bmatrix},$$

or

$$\{\ddot{q}\} = [\backslash\Omega^2\backslash]\{q\} = \frac{1}{J}\{Q\},$$

where $[\backslash\Omega^2\backslash]$ is the spectral matrix.

(x) *Example*:

Let us choose to exemplify $p^2 = \frac{k}{2J} = 100\,\mathrm{s}^{-2}$. It results for the spectral matrix:

$$[\backslash\Omega^2\backslash] = \begin{bmatrix} 0 & 0 & 0 \\ 0 & 14.14 & 0 \\ 0 & 0 & 26.45 \end{bmatrix}.$$

The system solution in canonical coordinates is:

$$\begin{Bmatrix} q_1 \\ q_1 \\ q_1 \end{Bmatrix} = \begin{Bmatrix} \dot{q}_{10}t + q_{10} \\ q_{20}\cos(14.14t + \varphi_{20}) \\ q_{30}\cos(26.45t + \varphi_{30}) \end{Bmatrix}.$$

The system solution in the coordinates of the equivalent model is:

$$\begin{Bmatrix}\theta_1\\\theta_2\\\theta_3\end{Bmatrix} = [\Phi_M]\{q\}$$

$$= \begin{Bmatrix}0.3785\\0.3785\\0.3785\end{Bmatrix}(\dot{q}_{10}t + q_{10}) + \begin{Bmatrix}0.365\\0\\-0.548\end{Bmatrix} q_{20}\cos(14.14t + \varphi_{20})$$

$$+ \begin{Bmatrix}0.239\\-0.597\\0.239\end{Bmatrix} q_{30}\cos(26.45t + \varphi_{30}).$$

The six integration constants are determined from the initial conditions:

$$\begin{Bmatrix}\theta_1\\\theta_2\\\theta_3\end{Bmatrix}_{t=0} = \begin{Bmatrix}\theta_{10}\\\theta_{20}\\\theta_{30}\end{Bmatrix}, \quad \begin{Bmatrix}\dot{\theta}_1\\\dot{\theta}_2\\\dot{\theta}_3\end{Bmatrix}_{t=0} = \begin{Bmatrix}\dot{\theta}_{10}\\\dot{\theta}_{20}\\\dot{\theta}_{30}\end{Bmatrix}.$$

It results the system:

$$\begin{aligned}
\theta_{10} &= 0.3785q_{10} + 0.365q_{20}\cos\varphi_{20} + 0.239q_{30}\cos\varphi_{30};\\
\theta_{20} &= 0.3785q_{10} - 0.597q_{30}\cos\varphi_{30};\\
\theta_{30} &= 0.3785q_{10} - 0.548q_{20}\cos\varphi_{20} + 0.239q_{30}\cos\varphi_{30},
\end{aligned}$$

and

$$\begin{aligned}
\dot{\theta}_{10} &= 0.3785\dot{q}_{10} - 5.161\dot{q}_{20}\sin\varphi_{20} + 6.322\dot{q}_{30}\sin\varphi_{30};\\
\dot{\theta}_{20} &= 0.3785\dot{q}_{10} + 15.791\dot{q}_{30}\sin\varphi_{30};\\
\dot{\theta}_{30} &= 0.3785\dot{q}_{10} + 7.749\dot{q}_{20}\sin\varphi_{20} + 6.322\dot{q}_{30}\sin\varphi_{30}.
\end{aligned}$$

From this system of six equations with six unknowns it results (integration constants) $q_{10}, \dot{q}_{10}, q_{20}, q_{30}, \varphi_{20}, \varphi_{30}$.

Loaded system

If on the system acts a constant torque $M_1 = 100$ N m with $J = 1$ kg m^2 but without a resistant torque, the motion equation becomes:

$$\{\ddot{q}\} + 100\begin{bmatrix}0 & 0 & 0\\0 & 2 & 0\\0 & 0 & 7\end{bmatrix}\{q\} = 100\begin{Bmatrix}0.3785\\0.365\\0.239\end{Bmatrix}.$$

Using the new vectorial function:

$$\{\hat{q}\} = \{q\} - \begin{bmatrix} 0 & 0 & 0 \\ 0 & \frac{1}{2} & 0 \\ 0 & 0 & \frac{1}{7} \end{bmatrix} \begin{Bmatrix} 0.3785 \\ 0.365 \\ 0.239 \end{Bmatrix} = \{q\} - \begin{Bmatrix} 0 \\ 0.1875 \\ 0.0341 \end{Bmatrix},$$

the system becomes:

$$\{\ddot{\hat{q}}\} + 100 \begin{bmatrix} 0 & 0 & 0 \\ 0 & 2 & 0 \\ 0 & 0 & 7 \end{bmatrix} \{\hat{q}\} = \begin{Bmatrix} 37.85 \\ 0 \\ 0 \end{Bmatrix},$$

with the solution:

$$\begin{aligned} \hat{q}_1 &= 37.85\frac{t^2}{2} + \dot{q}_{10}t + q_{10}, \\ \hat{q}_2 &= q_{20}\cos(\omega_2 t + \varphi_{20}), \\ \hat{q}_3 &= q_{30}\cos(\omega_3 t + \varphi_{30}), \end{aligned}$$

The first solution is the rigid motion which, in the absence of a resistant moment, is uniformly accelerated and the other two are harmonic oscillations whose overlapping provides the wheels motion. Thus, it can be written:

$$\begin{Bmatrix} \theta_1 \\ \theta_2 \\ \theta_3 \end{Bmatrix} = [\Phi_M] \begin{Bmatrix} 37.85\frac{t^2}{2} + \dot{q}_{10}t + q_{10} \\ q_{20}\cos(\omega_2 t + \varphi_{20}) + 0.1875 \\ q_{30}\cos(\omega_3 t + \varphi_{30}) + 0.0341 \end{Bmatrix}.$$

(xi) *Transmission of torques*

Let us consider the transmission of a torque of the form:

$$\{M_1\} = \begin{Bmatrix} 100 \\ 0 \\ 0 \end{Bmatrix} \cos 50\,t$$

and a resistant torque of the form:

$$\{M_2'\} = \begin{Bmatrix} 0 \\ 0 \\ 100 \end{Bmatrix} \cos 10\,t.$$

Let us consider the system of inertial normalized equations:

$$\{\ddot{q}\} + 100\begin{bmatrix} 0 & 0 & 0 \\ 0 & 2 & 0 \\ 0 & 0 & 7 \end{bmatrix}\{q\} = 100\begin{Bmatrix} 0.3785 \\ 0.365 \\ 0.239 \end{Bmatrix}\cos 50t + 100\begin{Bmatrix} 0.3785 \\ -0.548 \\ 0.239 \end{Bmatrix}\cos 10t,$$

with the particular forced solution:

$$\begin{Bmatrix} q_1 \\ q_2 \\ q_3 \end{Bmatrix}_p = \begin{bmatrix} -\frac{1}{25} & 0 & 0 \\ 0 & \frac{1}{23} & 0 \\ 0 & 0 & -\frac{1}{18} \end{bmatrix}\begin{Bmatrix} 0.3785 \\ 0.365 \\ 0.239 \end{Bmatrix}\cos 50t + \begin{bmatrix} -1 & 0 & 0 \\ 0 & 1 & 0 \\ 0 & 0 & -\frac{1}{6} \end{bmatrix}\begin{Bmatrix} 0.3785 \\ -0.548 \\ 0.239 \end{Bmatrix}\cos 10t.$$

This solution will be added to the homogeneous system solution previously determined. The integration constants shall be determined for this solution (composed by the homogeneous solution and the particular solution).

(xii) *Rayleigh's quotient*

We will approximate the eigenvalues using the *Rayleigh's quotient*. For the first mode of motion we choose the eigenvector as:

$$\{\Phi_1\}^T = \begin{bmatrix} 1 & 1 & 1 \end{bmatrix}.$$

The first eigenvalue follows:

$$\omega_1^2 = \frac{\{\Phi_1\}^T[K]\{\Phi_1\}}{\{\Phi_1\}^T[M]\{\Phi_1\}} = 0.$$

For the second eigenmode we choose as:

$$\{\Phi_2\}^T = \begin{bmatrix} 1 & 0 & -1 \end{bmatrix},$$

and the eigenvalue derives from this relationship:

$$\omega_2^2 = \frac{\{\Phi_2\}^T[K]\{\Phi_2\}}{\{\Phi_2\}^T[M]\{\Phi_2\}} = \frac{k}{J}\frac{\begin{bmatrix} 1 & 0 & -1 \end{bmatrix}\begin{bmatrix} 3 & -3 & 0 \\ -3 & 5 & -2 \\ 0 & -2 & 2 \end{bmatrix}\begin{Bmatrix} 1 \\ 0 \\ -1 \end{Bmatrix}}{\begin{bmatrix} 1 & 0 & -1 \end{bmatrix}\begin{bmatrix} 3 & 0 & 0 \\ 0 & 2 & 0 \\ 0 & 0 & 2 \end{bmatrix}\begin{Bmatrix} 1 \\ 0 \\ -1 \end{Bmatrix}} = \frac{k}{J} = 200,$$

from which $\omega_2 = 14.14\,\mathrm{s}^{-1}$, which coincides with the value obtained by exactly calculating the result. To calculate the third eigenvalue, the vector will be chosen as:

$$\{\Phi_3\}^T = \begin{bmatrix} 1 & -2 & 1 \end{bmatrix}.$$

It results:

$$\omega_3^2 = \frac{\{\Phi_3\}^T[K]\{\Phi_3\}}{\{\Phi_3\}^T[M]\{\Phi_3\}} = \frac{k}{J}\frac{\begin{bmatrix}1 & -2 & 1\end{bmatrix}\begin{bmatrix}3 & -3 & 0\\ -3 & 5 & -2\\ 0 & -2 & 2\end{bmatrix}\begin{Bmatrix}1\\ -2\\ 1\end{Bmatrix}}{\begin{bmatrix}1 & -2 & 1\end{bmatrix}\begin{bmatrix}3 & 0 & 0\\ 0 & 2 & 0\\ 0 & 0 & 2\end{bmatrix}\begin{Bmatrix}1\\ -2\\ 1\end{Bmatrix}} = \frac{45k}{13J} \cong 692,$$

thus

$$\omega_3 = 26.31\,\mathrm{s}^{-1},$$

against the exact result of 26.45 Hz. Thus, we obtain in this case a good approximation.

(xiii) *Transforming the coordinates*

It is natural for the study of the system to introduce a coordinate α_1 that describes the rigid motion of the system. We will force this coordinate to respect the condition that the total moment of momentum of the system is equal to the moment of momentum of the transmission considered as rigid and rotating with the angle α_1. Thus:

$$J_1\dot{\varphi}_1 + \frac{J_2}{i_2}\dot{\varphi}_2 + \frac{J_3}{i_3}\dot{\varphi}_3 = J\dot{\alpha}_1,$$

from where:

$$J = J_1 + \frac{J_2}{i_2^2} + \frac{J_3}{i_3^2},$$

considering that

$$\begin{Bmatrix}\varphi_1\\ \varphi_2\\ \varphi_3\end{Bmatrix} = \begin{Bmatrix}1\\ \frac{1}{i_2}\\ \frac{1}{i_3}\end{Bmatrix}\alpha_1.$$

Also, coordinates α_2 and α_3 will be introduced and represent the relative motion of the wheels relative to the rigid motion α_1:

$$\alpha_2 = \varphi_1 - i_2\varphi_2,$$

$$\alpha_3 = \varphi_2 - \frac{i_3}{i_2}\varphi_3.$$

Thus:

$$\begin{Bmatrix} \alpha_1 \\ \alpha_2 \\ \alpha_3 \end{Bmatrix} = \begin{bmatrix} J_1' & \frac{J_2'}{i_2} & \frac{J_3'}{i_3} \\ 1 & -i_2 & 0 \\ 0 & 1 & -\frac{i_3}{i_2} \end{bmatrix} \begin{Bmatrix} \varphi_1 \\ \varphi_2 \\ \varphi_3 \end{Bmatrix},$$

with: $J_1' = \frac{J_1}{J}$, $J_2' = \frac{J_2}{J}$, $J_3' = \frac{J_3}{J}$.

It results:

$$\begin{Bmatrix} \varphi_1 \\ \varphi_2 \\ \varphi_3 \end{Bmatrix} = \begin{bmatrix} J_1' & \frac{J_2'}{i_2^2} + \frac{J_3'}{i_3^2} & \frac{J_3' i_2}{i_3^2} \\ \frac{1}{i_2} & -\frac{J_1'}{i_2} & \frac{J_3'}{i_3^2} \\ \frac{1}{i_3} & -\frac{J_1'}{i_3} & -\frac{J_1' i_2}{i_3^2} - \frac{J_3'}{i_2 i_3} \end{bmatrix} \begin{Bmatrix} \alpha_1 \\ \alpha_2 \\ \alpha_3 \end{Bmatrix} = [L] \begin{Bmatrix} \alpha_1 \\ \alpha_2 \\ \alpha_3 \end{Bmatrix}.$$

If the calculus are made:

$$[L]^T [J][L] = \begin{bmatrix} J & 0 & 0 \\ 0 & \frac{J_1 (J_2^* + J_3^*)}{J} & \frac{J_1 J_3^*}{J} \\ 0 & \frac{J_1 J_3^*}{J} & \frac{J_3 (J_1^* + J_2^*)}{J} \end{bmatrix},$$

with $J_2^* = \frac{J_2}{i_2^2}$ and $J_3^* = \frac{J_3}{i_3^2}$, and:

$$[L]^T [K][L] = \begin{bmatrix} 0 & 0 & 0 \\ 0 & k_1 & 0 \\ 0 & 0 & k_2^* \end{bmatrix},$$

where $k_2^* = \frac{k_2}{i_2^2}$.

The motion equations become:

$$J \ddot{\alpha}_1 = 0,$$

$$\frac{1}{J} \begin{bmatrix} J_1 (J_2^* + J_3^*) & J_1 J_3^* \\ J_1 J_3^* & J_3 (J_1^* + J_2^*) \end{bmatrix} \begin{Bmatrix} \ddot{\alpha}_2 \\ \ddot{\alpha}_3 \end{Bmatrix} + \begin{bmatrix} k_1 & 0 \\ 0 & k_2^* \end{bmatrix} \begin{Bmatrix} \alpha_2 \\ \alpha_3 \end{Bmatrix} = \{0\}.$$

The last two equations are inertial (dynamic) coupled and elastic (static) decoupled.

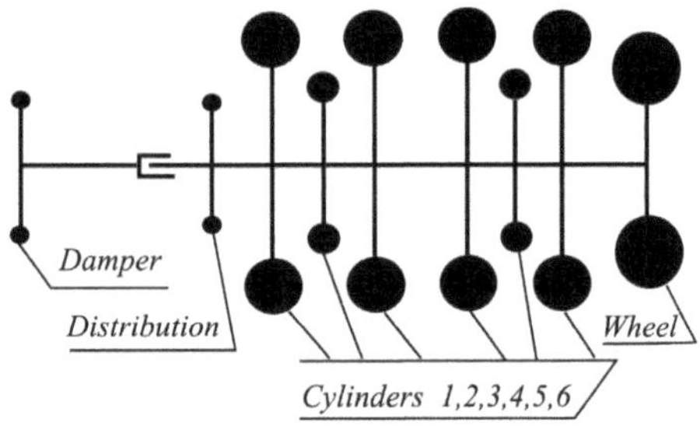

Fig. 6.6 The model of a crankshaft

A2 A crankshaft of an engine is considered, which has at one end a viscous damper. The system can be modeled as a system with concentrated masses (flywheels) as shown in Fig. 6.6.

The equations of undamped free vibration of the system are:

$$[J]\{\ddot{\varphi}\} + [K]\{\varphi\} = \{0\},$$

where:

$$[J] = \begin{bmatrix} J_1 & & & & & & & & \\ & J_2 & & & & & & & \\ & & J_3 & & & 0 & & & \\ & & & J_4 & & & & & \\ & & & & J_5 & & & & \\ & & 0 & & & J_6 & & & \\ & & & & & & J_7 & & \\ & & & & & & & J_8 & \\ & & & & & & & & J_9 \end{bmatrix},$$

$$[K] = \begin{bmatrix} 0 & & & & & & & & \\ & k_1 & -k_1 & & & & & & \\ & -k_1 & k_1+k_2 & -k_2 & & & 0 & & \\ & & -k_2 & k_2+k_3 & -k_3 & & & & \\ & & & -k_3 & k_3+k_4 & -k_4 & & & \\ & & & & -k_4 & k_4+k_5 & -k_5 & & \\ & & 0 & & & -k_5 & k_5+k_6 & -k_6 & \\ & & & & & & -k_6 & k_6+k_7 & -k_7 \\ & & & & & & & -k_7 & k_7 \end{bmatrix},$$

$$\{\varphi\}^T = \begin{bmatrix} \varphi_1 & \varphi_2 & \varphi_3 & \varphi_4 & \varphi_5 & \varphi_6 & \varphi_7 & \varphi_8 & \varphi_9 \end{bmatrix}.$$

The characteristic equation

$$\det\left([K]-\omega^2[M]\right)=0,$$

gives the circular eigenfrequencies in the absence of damping.

The mechanical constants appearing in the array can be computed using classical relationships (Harris and Crede 1961). If the influence of vibration damping is also considered, the free vibration motion equations are:

$$[J][\ddot{\varphi}]+[C][\dot{\varphi}]+[K][\varphi]=\{0\},$$

with:

$$[C]=\begin{bmatrix} c & -c & & & & & & & \\ -c & c & & & & & & & \\ & & 0 & & & & 0 & & \\ & & & 0 & & & & & \\ & & & & 0 & & & & \\ & & 0 & & & 0 & & & \\ & & & & & & 0 & & \\ & & & & & & & 0 & \\ & & & & & & & & 0 \end{bmatrix}.$$

The characteristic equation

$$\det\left([K]-\omega^2[M]\right)=0,$$

gives the circular eigenfrequencies in the absence of damping.

The linear homogeneous system

$$\left([K]-\omega_i^2[M]\right)\{\Phi_i\}=\{0\},$$

offers the eigenmodes represented in Fig. 6.7, i.e. for an engine whose mechanical characteristics have been calculated. It can be noted that the vibration modes of the shaft are decoupled from the damping mode of motion (the second rigid motion mode, as well as the first and the second eigenvalues are zeros).

A3 Coupled harmonic oscillators. Consider two pendulums, which are elastically coupled (Fig. 6.8). The motion equations will be:

$$\begin{aligned} J_1\ddot{\theta}_1 &= -m_1 gL\sin\theta_1+ka^2(\theta_2-\theta_1),\\ J_2\ddot{\theta}_2 &= -m_2 gL\sin\theta_1+ka^2(\theta_2-\theta_1). \end{aligned}$$

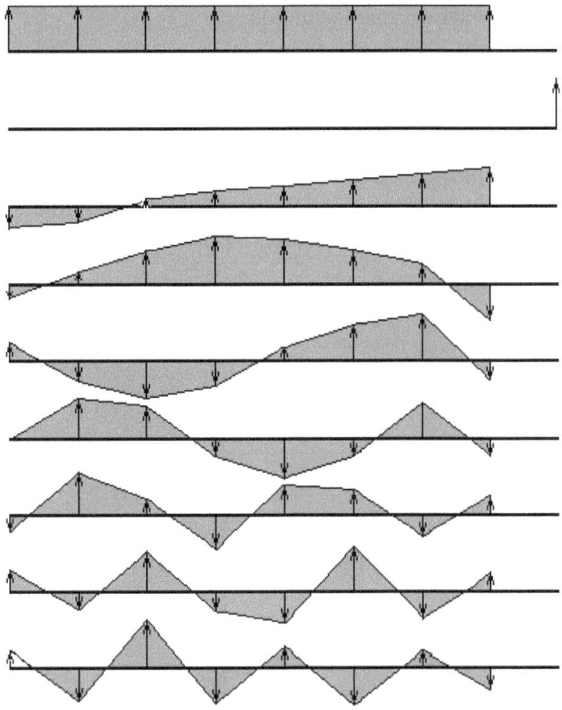

Fig. 6.7 Eigenmodes of vibration of the crankshaft

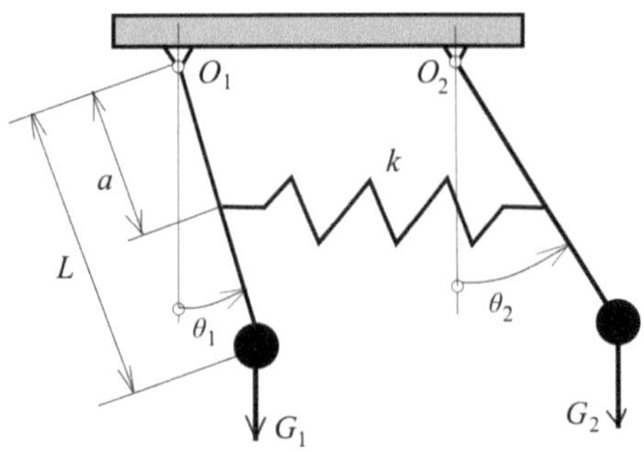

Fig. 6.8 Coupled pendulum

For small oscillations we can consider that $\sin\theta_1 \cong \theta_1$ and $\sin\theta_2 \cong \theta_2$. By neglecting the masses of the two bars, we also can obtain: $J_1 = m_1 L^2$ and $J_2 = m_2 L^2$. Under these assumptions the motion equations become:

$$L^2 \begin{bmatrix} m_1 & 0 \\ 0 & m_2 \end{bmatrix} \begin{Bmatrix} \ddot{\theta}_1 \\ \ddot{\theta}_2 \end{Bmatrix} + \begin{bmatrix} ka^2 + m_1 gL & -ka^2 \\ -ka^2 & ka^2 + m_2 gL \end{bmatrix} \begin{Bmatrix} \theta_1 \\ \theta_2 \end{Bmatrix} = \begin{Bmatrix} 0 \\ 0 \end{Bmatrix}.$$

To simplify the calculus, consider: $m_1 = m_2 = m$. Let us note further: $p_o^2 = \frac{g}{L}$; $p_1^2 = \frac{ka^2}{mL^2}$. With this notation, the motion equations can be written:

$$\{\ddot{\theta}\} + \begin{bmatrix} p_o^2 + p_1^2 & -p_1^2 \\ -p_1^2 & p_o^2 + p_1^2 \end{bmatrix} \{\theta\} = \{0\}.$$

The characteristic equation

$$\begin{vmatrix} p_o^2 + p_1^2 - \omega^2 & -p_1^2 \\ -p_1^2 & p_o^2 + p_1^2 - \omega^2 \end{vmatrix} = 0,$$

gives the eigenfrequencies of the system.

$$\omega_1^2 = p_o^2; \quad \omega_2^2 = p_o^2 + 2p_1^2.$$

For the first eigenvalue, the first eigenvector is given by the homogeneous system:

$$\begin{bmatrix} p_1^2 & -p_1^2 \\ -p_1^2 & p_1^2 \end{bmatrix} \begin{Bmatrix} e_1 \\ e_2 \end{Bmatrix} = \begin{Bmatrix} 0 \\ 0 \end{Bmatrix};$$

where, if one chooses $e_1 = 1$, one gets:

$$\begin{Bmatrix} e_1 \\ e_2 \end{Bmatrix} = \begin{Bmatrix} 1 \\ 1 \end{Bmatrix}.$$

This eigenmode is a motion of the system as if the two pendulums were connected by a rigid bar, articulated at the ends and vibrating with the eigenfrequency $p_o = \sqrt{g/L}$ (as a single pendulum).

For the second eigenvalue we obtain:

$$\begin{bmatrix} -p_1^2 & -p_1^2 \\ -p_1^2 & -p_1^2 \end{bmatrix} \begin{Bmatrix} e_1 \\ e_2 \end{Bmatrix} = \begin{Bmatrix} 0 \\ 0 \end{Bmatrix},$$

with the solution:

$$\begin{Bmatrix} e_1 \\ e_2 \end{Bmatrix} = \begin{Bmatrix} 1 \\ -1 \end{Bmatrix}.$$

It is a "symmetrical" mode, i.e. the two pendulums will oscillate, with the same eigenfrequency, but in the opposite direction (the position of one is obtain d from the position of the other one by mirroring).

The general solution becomes:

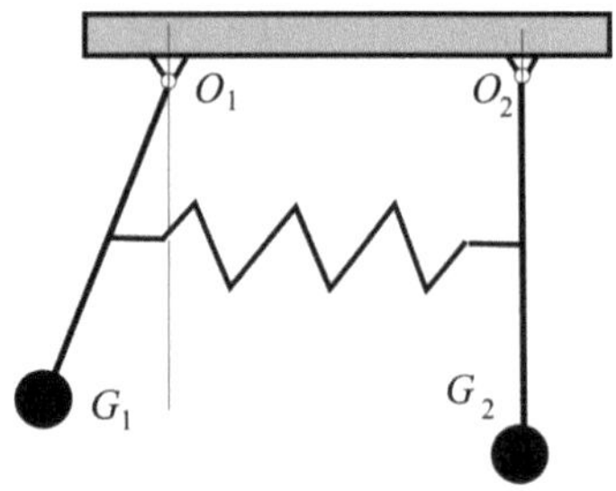

Fig. 6.9 Initial conditions

$$\begin{Bmatrix} \theta_1 \\ \theta_2 \end{Bmatrix} = A_1 \begin{Bmatrix} 1 \\ 1 \end{Bmatrix} \cos(\omega_1 t + \varphi_1) + A_2 \begin{Bmatrix} 1 \\ -1 \end{Bmatrix} \cos(\omega_2 t + \varphi_2).$$

The integration constants $A_1, A_2, \varphi_1, \varphi_2$ result from the boundary conditions:

$$\begin{Bmatrix} \theta_1 \\ \theta_2 \end{Bmatrix}_{|t=0} = \begin{Bmatrix} \theta_{10} \\ \theta_{20} \end{Bmatrix}, \begin{Bmatrix} \dot{\theta}_1 \\ \dot{\theta}_2 \end{Bmatrix}_{|t=0} = \begin{Bmatrix} \dot{\theta}_{10} \\ \dot{\theta}_{20} \end{Bmatrix} = \begin{Bmatrix} \omega_{10} \\ \omega_{20} \end{Bmatrix},$$

which leads to the following linear system:

$$A_1 \begin{Bmatrix} 1 \\ 1 \end{Bmatrix} \cos\varphi_1 + A_2 \begin{Bmatrix} 1 \\ -1 \end{Bmatrix} \cos\varphi_2 = \begin{Bmatrix} \theta_{10} \\ \theta_{20} \end{Bmatrix},$$

$$A_1\omega_1 \begin{Bmatrix} 1 \\ 1 \end{Bmatrix} \sin\varphi_1 + A_2\omega_2 \sin\varphi_2 \begin{Bmatrix} 1 \\ -1 \end{Bmatrix} \sin\varphi_2 = -\begin{Bmatrix} \omega_{10} \\ \omega_{20} \end{Bmatrix}.$$

To simplify calculus, consider the particular case:

$$\begin{Bmatrix} \theta_{10} \\ \theta_{20} \end{Bmatrix} = \begin{Bmatrix} 2a \\ 0 \end{Bmatrix} \text{ and } \begin{Bmatrix} \omega_{10} \\ \omega_{20} \end{Bmatrix} = \begin{Bmatrix} 0 \\ 0 \end{Bmatrix},$$

i.e. the system starts from rest as in Fig. 6.9. It will be obtained: $A_1 = A_2 = a$ and $\varphi_1 = \varphi_2 = 0$. Under these conditions, the system response is:

$$\begin{Bmatrix} \theta_1 \\ \theta_2 \end{Bmatrix} = a\left(\begin{Bmatrix} 1 \\ 1 \end{Bmatrix} \cos\omega_1 t + \begin{Bmatrix} 1 \\ -1 \end{Bmatrix} \cos\omega_2 t\right)$$
$$= 2a \begin{Bmatrix} \cos\frac{\omega_1+\omega_2}{2}t \cos\frac{\omega_2-\omega_1}{2}t \\ \sin\frac{\omega_1+\omega_2}{2}t \sin\frac{\omega_2-\omega_1}{2}t \end{Bmatrix}.$$

Beats. In the case when $\frac{ka^2}{mL^2} \ll \frac{g}{L}$ (and when two frequencies get close) the beating phenomenon occurs. When θ_1 has the maximum amplitude on the modulation characteristic, θ_2 has the value zero and reciprocal.

If we denote:

$$\sigma = \frac{\omega_1 + \omega_2}{2} = \frac{p_o + \sqrt{p_o^2 + 2p_1^2}}{2} \cong p_o,$$

$$\delta = \frac{\omega_1 - \omega_2}{2} = \frac{-p_o + \sqrt{p_o^2 + 2p_1^2}}{2} = \frac{-p_o^2 + p_o^2 + 2p_1^2}{2\left(p_o + \sqrt{p_o^2 + 2p_1^2}\right)} \cong \frac{p_1^2}{2p_o},$$

it can be written:

$$\begin{Bmatrix} \theta_1 \\ \theta_2 \end{Bmatrix} = \begin{Bmatrix} 2a \cos \sigma t \cos \delta t \\ 2a \sin \sigma t \sin \delta t \end{Bmatrix} = \begin{Bmatrix} A^*(t) \cos \sigma t \\ B^*(t) \sin \sigma t \end{Bmatrix},$$

where $A^*(t) = 2a \cos \delta t$; $B^*(t) = 2a \sin \delta t$ are the variable amplitudes in time (modulation) of the oscillations with the eigenpulsation δ. The pseudo-period of the oscillating phenomenon is: $T_\sigma = \frac{2\pi}{p_o}$. $T_\delta = \frac{4\pi p_o}{p_1^2}$ is the period of the modulating one. Periods will be in the ratio:

$$\frac{T_\sigma}{T_\delta} = \frac{1}{2}\left(\frac{p_1}{p_o}\right)^2,$$

which ensures that the two characteristics are well separated. Therefore, the weak elastic coupling realizes a transfer of energy between the two pendulums with the period T_δ, the energy of the system being preserved on the whole (Fig. 6.10).

A4 Consider the model of a truck. We want to study the vibrations of the sprung and unsprung masses of the truck. The dimensions of the system are given in Fig. 6.11.

We will consider the generalized independent coordinates $\{q\} = \left[q_1\ q_2\ \cdots\ q_7\right]^T$ defined as follows (Tofan and Vlase 1985):

$$\begin{aligned} &q_1 = x, \quad q_1 = x', \quad q_3 = x'', \quad q_4 = \alpha; \\ &q_5 = \beta, \quad q_6 = \beta', \quad q_7 = \beta''. \end{aligned}$$

The kinematic system will give us the relationships that exist between the coordinates of the interesting points, written according to the independent generalized coordinates:

$$x_1 = q_1 + \frac{A}{2}q_4 + \frac{E}{2}q_5, \quad x_5 = q_2 + \frac{E}{2}q_6,$$

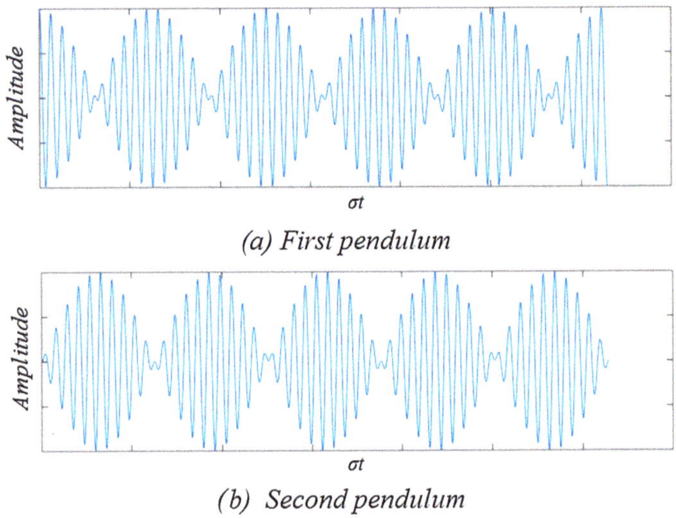

Fig. 6.10 Beats. The amplitudes of the two pendulums(qualitative representations)

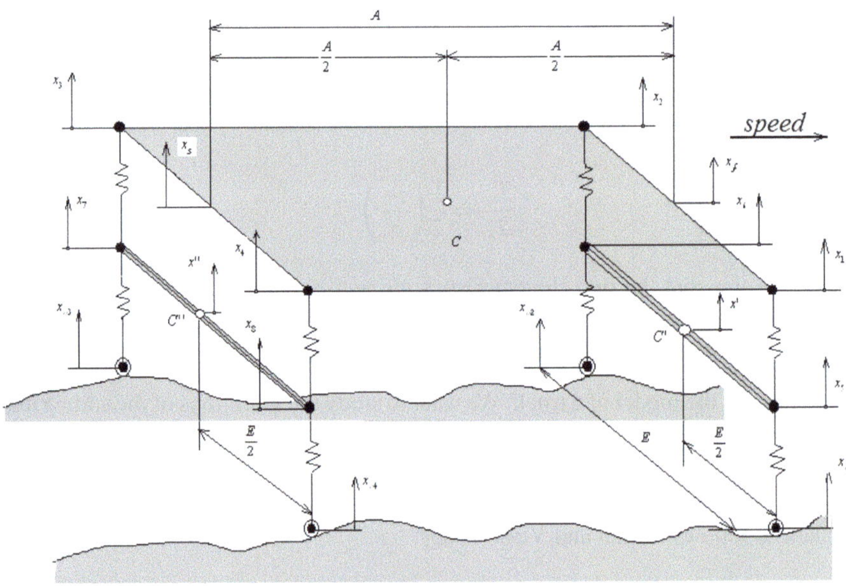

Fig. 6.11 Mathematical model of a truck for the study of transverse vibrations

$$x_2 = q_1 + \frac{A}{2}q_4 - \frac{E}{2}q_5, \quad x_6 = q_2 - \frac{E}{2}q_6,$$
$$x_3 = q_1 - \frac{A}{2}q_4 + \frac{E}{2}q_5, \quad x_7 = q_3 + \frac{E}{2}q_7,$$

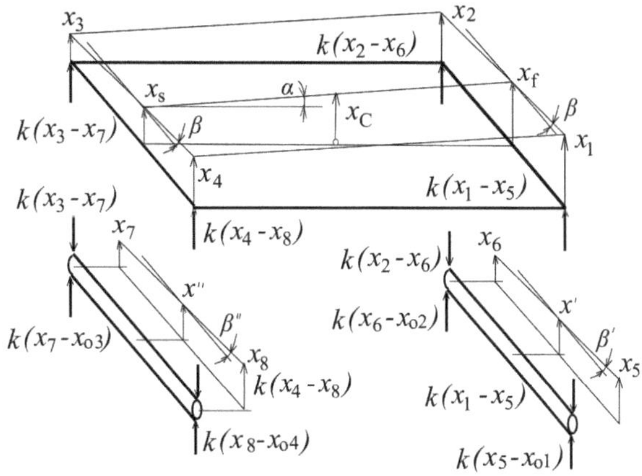

Fig. 6.12 Free body diagram

$$x_4 = q_1 - \frac{A}{2}q_4 - \frac{E}{2}q_5, \quad x_8 = q_3 - \frac{E}{2}q_7.$$

The forces that occur in the suspension springs and the forces in the tires are:

$$\begin{aligned}
f_1' &= -k(x_1 - x_5), \quad f_2' = -k(x_2 - x_6),\\
f_3' &= -k(x_3 - x_7), \quad f_4' = -k'(x_4 - x_8),\\
f_5' &= -k(x_5 - x_{01}), \quad f_6' = -k(x_6 - x_{02}),\\
f_7' &= -k(x_7 - x_{03}), \quad f_8' = -k(x_8 - x_{04}).
\end{aligned}$$

We can write the system as:

$$\{x\} = [A]\{q\},$$

where the matrix $[A]$ is given by (Fig. 6.12):

$$[A] = \begin{bmatrix}
1\,0\,0 & -\frac{A}{2} & \frac{E}{2} & 0 & 0\\
1\,0\,0 & -\frac{A}{2} & -\frac{E}{2} & 0 & 0\\
1\,0\,0 & \frac{A}{2} & \frac{E}{2} & 0 & 0\\
1\,0\,0 & \frac{A}{2} & -\frac{E}{2} & 0 & 0\\
0\,1\,0 & 0 & 0 & \frac{E}{2} & 0\\
0\,1\,0 & 0 & 0 & -\frac{E}{2} & 0\\
0\,0\,1 & 0 & 0 & 0 & \frac{E}{2}\\
0\,0\,1 & 0 & 0 & 0 & -\frac{E}{2}
\end{bmatrix}.$$

The motion equations for the system components are given by:

$$
\begin{aligned}
&m\ddot{x}_C + k(x_1 - x_5) + k(x_2 - x_6) + k(x_3 - x_7) + k(x_4 - x_8) = 0,\\
&m'\ddot{x}' - k(x_1 - x_5) - k(x_2 - x_6) + k'x_5 + k'x_6 = 0,\\
&m''\ddot{x}'' - k(x_3 - x_7) - k(x_4 - x_8) + k'x_7 + k'x_8 = 0,
\end{aligned}
$$

$$
\begin{aligned}
&J_\alpha\ddot{\alpha} + k(x_1 - x_5)\frac{A}{2} + k(x_2 - x_6)\frac{A}{2} - k(x_3 - x_7)\frac{A}{2} - k(x_4 - x_8)\frac{A}{2} = 0,\\
&J_\beta\ddot{\beta} + k(x_1 - x_5)\frac{E}{2} - k(x_2 - x_6)\frac{E}{2} - k(x_3 - x_7)\frac{E}{2} + k(x_4 - x_8)\frac{E}{2} = 0,\\
&J_{\beta'}\ddot{\beta}' + [k(x_5 - x_1) - k'x_5]\frac{E}{2} - [k(x_6 - x_2) - k'x_6]\frac{E}{2} = 0,\\
&J_{\beta''}\ddot{\beta}'' + [k(x_8 - x_4) - k'x_8]\frac{E}{2} - [k(x_7 - x_3) - k'x_7]\frac{E}{2} = 0.
\end{aligned}
$$

Taking into account the above written kinematic conditions will result in:

$$
\begin{aligned}
&m\ddot{x}_C + 4kx_c - 2kx' - 2kx'' = 0,\\
&m'\ddot{x}' + 2(k + k')x' + k(2x_c + A\alpha) = 0,\\
&m'\ddot{x}'' + 2(k + k')x'' + k(2x_c - A\alpha) = 0,
\end{aligned}
$$

$$
\begin{aligned}
&J_\alpha\ddot{\alpha} + kA(A\alpha - x' + x'') = 0,\\
&J_\beta\ddot{\beta} + kE(E\beta - E\beta' - E\beta'') = 0,\\
&J_{\beta'}\ddot{\beta}' + (k + k')\frac{E}{2}E\beta' - k\frac{E^2}{2}\beta = 0,\\
&J_{\beta''}\ddot{\beta}'' + (k + k')\frac{E}{2}E\beta'' + k(2x_C + A\alpha)\frac{E}{2} = 0.
\end{aligned}
$$

If we use the notations:

$$
[M] = \begin{bmatrix}
m & 0 & 0 & 0 & 0 & 0 & 0\\
0 & m' & 0 & 0 & 0 & 0 & 0\\
0 & 0 & m'' & 0 & 0 & 0 & 0\\
0 & 0 & 0 & J_\alpha & 0 & 0 & 0\\
0 & 0 & 0 & 0 & J_\beta & 0 & 0\\
0 & 0 & 0 & 0 & 0 & J_{\beta'} & 0\\
0 & 0 & 0 & 0 & 0 & 0 & J_{\beta''}
\end{bmatrix},
$$

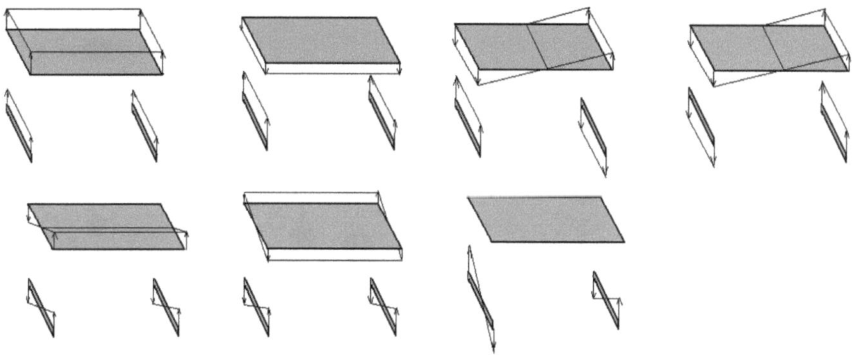

Fig. 6.13 Eigenmodes of vibration

$$[K] = \begin{bmatrix} 4k & -2k & -2k & 0 & 0 & 0 & 0 \\ -2k & 2(k+k') & 0 & -kA & 0 & 0 & 0 \\ -2k & 0 & 2(k+k') & +kA & 0 & 0 & 0 \\ 0 & -kA & kA & kA^2 & 0 & 0 & 0 \\ 0 & 0 & 0 & 0 & kE^2 & -k\frac{E^2}{2} & -k\frac{E^2}{2} \\ 0 & 0 & 0 & 0 & -k\frac{E^2}{2} & (k+k')\frac{E^2}{2} & 0 \\ 0 & 0 & 0 & 0 & -k\frac{E^2}{2} & 0 & (k+k')\frac{E^2}{2} \end{bmatrix}$$

then the differential equations of the free vibrations of the system are given by:

$$[M]\{\ddot{q}\} + [K]\{q\} = \{0\}.$$

For a real case the modal matrix is:

$$[\Phi] = \begin{bmatrix} 0.7132 & -0.0978 & 0 & 0 & 0 & 0 & 0 \\ 0.4956 & 0.7037 & -0.7022 & 0.6605 & 0 & 0 & 0 \\ 0.4956 & 0.7037 & 0.7022 & -0.6605 & 0 & 0 & 0 \\ 0 & 0 & 0.1180 & 0.3572 & 0 & 0 & 0 \\ 0 & 0 & 0 & 0 & 0.7132 & -0.0978 & 0 \\ 0 & 0 & 0 & 0 & 0.4956 & 0.7037 & -0.7071 \\ 0 & 0 & 0 & 0 & 0.4956 & 0.7037 & 0.7071 \end{bmatrix}$$

Furthermore (see Fig. 6.13), the vibration modes of the masses and unsprung masses of the vehicle will be represented. In this case, the degree of symmetry is low and occurs only in the last mode of vibration that is identical to that of the unsprung mass (beam axle), if studied separately.

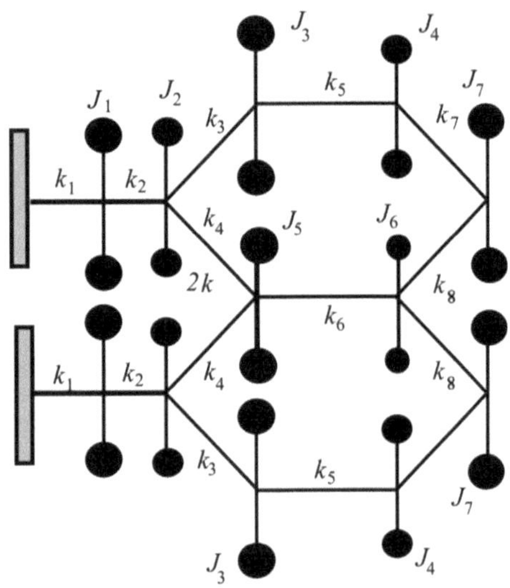

Fig. 6.14 A symmetrical system

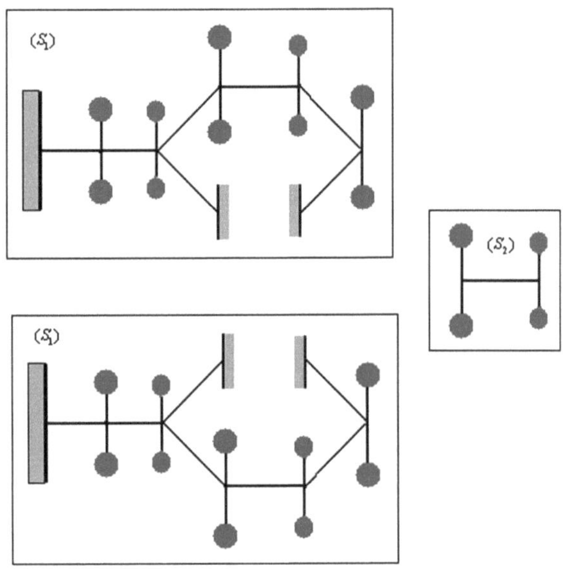

Fig. 6.15 Descomposing in subsystems

A5 Consider the system with symmetries presented in Fig. 6.14. As shown in Fig. 6.15, two identical branches can be isolated. We consider the following values for the moments of inertia $J_1 = J$, $J_2 = 2J$, $J_3 = 4J$, $J_4 = J$, $J_5 = 3J$, $J_6 = 7J$, and for the elastic constants $k_1 = k$, $k_2 = 3k$, $k_3 = 5k$, $k_4 = 3k$, $k_5 = 4k$, $k_6 = 7k$, $k_7 = k$, $k_8 = 3k$.

If we introduce the ratio $p^2 = k/J$, the eigenfrequencies are: $2853p$, $2754p$, $2438p$, $2415p$, $2181p$, $1691p$, $1667p$, $1212p$, $917p$, $780p$, $659p$, $194p$.

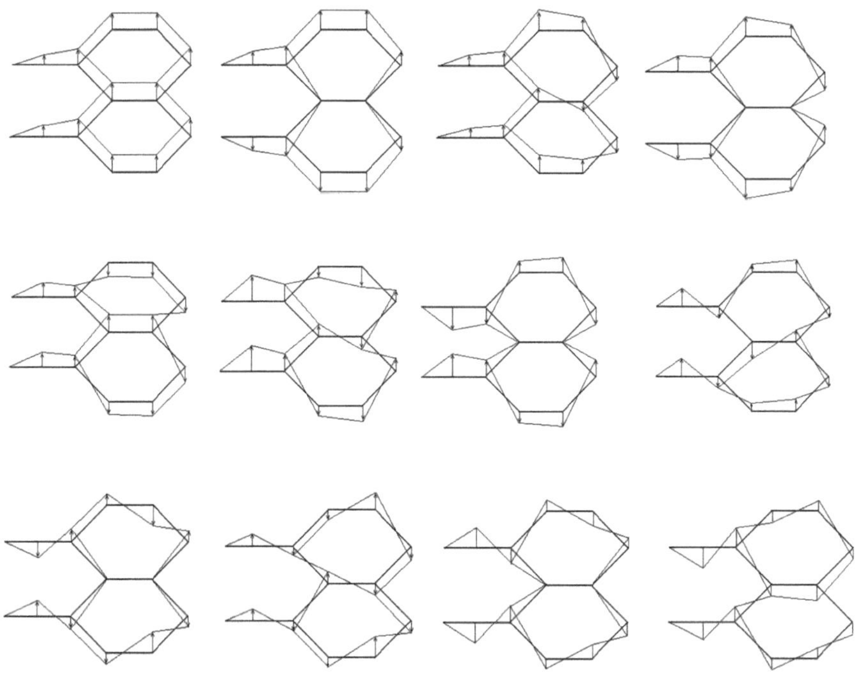

Fig. 6.16 Eigenmodes of vibration

The system is symmetric and can be decomposed into subsystems as shown in Fig. 6.15 (Mangeron et al. 1991; Zingoni 2005; Singer 2004). We find that we have two identical subsystems.

The modal matrix results after calculus:

$$\begin{bmatrix}
-0.37 & 0.41 & 0.13 & -0.24 & 0.51 & -0.62 & 0.59 & 0.15 & 0.20 & 0.25 & 0.22 & 0.21 \\
-0.37 & -0.41 & 0.13 & 0.24 & 0.51 & 0.62 & 0.59 & 0.15 & -0.20 & 0.25 & -0.22 & 0.21 \\
0.51 & -0.49 & -0.08 & 0.15 & -0.13 & -0.23 & 0.24 & 0.13 & 0.21 & 0.29 & 0.27 & 0.27 \\
0.51 & 0.49 & -0.08 & -0.15 & -0.13 & 0.23 & 0.24 & 0.13 & -0.21 & 0.29 & -0.27 & 0.27 \\
-0.14 & 0.16 & -0.15 & 0.13 & 0.02 & 0.12 & -0.14 & -0.17 & 0.27 & 0.39 & 0.41 & 0.29 \\
0.18 & -0.25 & 0.66 & -0.63 & 0.12 & 0.22 & -0.23 & -0.29 & 0.12 & 0.28 & 0.40 & 0.30 \\
-0.01 & 0.01 & -0.02 & 0.03 & -0.04 & -0.02 & 0.07 & -0.34 & -0.58 & -0.32 & 0.22 & 0.33 \\
-0.30 & 0.00 & 0.15 & 0.00 & -0.60 & 0.00 & 0.08 & 0.48 & 0.00 & 0.03 & 0.00 & 0.30 \\
0.05 & 0.00 & -0.03 & 0.00 & 0.22 & 0.00 & -0.15 & 0.48 & 0.00 & -0.20 & 0.00 & 0.32 \\
-0.14 & -0.16 & -0.15 & -0.03 & 0.02 & -0.12 & -0.14 & -0.17 & -0.27 & 0.39 & -0.41 & 0.29 \\
0.18 & 0.25 & 0.66 & 0.63 & 0.12 & -0.22 & -0.23 & -0.29 & -0.12 & 0.28 & -0.40 & 0.30 \\
-0.01 & -0.01 & -0.02 & -0.03 & -0.04 & 0.02 & 0.07 & -0.34 & 0.58 & -0.32 & -0.22 & 0.33
\end{bmatrix}.$$

The eigenvectors are presented in Fig. 6.16.

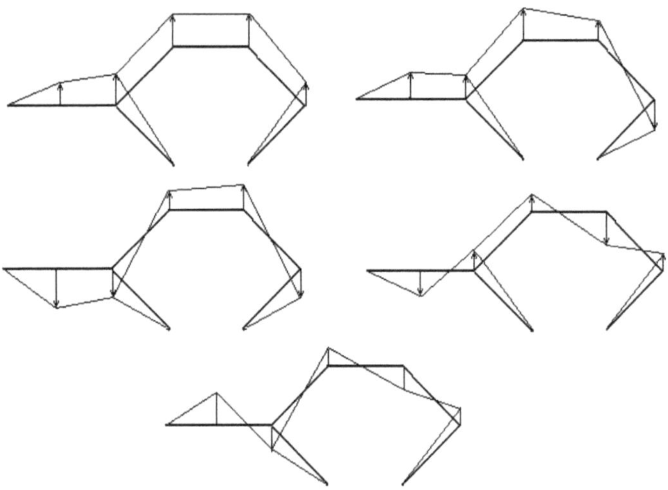

Fig. 6.17 Eigenmodes of the symmetrical part of the system

For the subsystem from Fig. 6.15, if the calculation of the eigenvalues is made, we obtain the spectrum $2754p$, $2415p$, $1691p$, $917p$, $659p$ and the corresponding modal matrix will be:

$$[\Phi_1] = \begin{bmatrix} -0.58 & 0.34 & 0.87 & -0.30 & 0.36 \\ 0.69 & -0.21 & 0.33 & -0.18 & 0.41 \\ -0.23 & -0.18 & -0.18 & -0.07 & 0.62 \\ 0.36 & 0.90 & -0.30 & 0.22 & 0.56 \\ -0.01 & -0.05 & 0.09 & 0.91 & 0.07 \end{bmatrix}.$$

The eigenmodes of vibration in this case are represented in Fig. 6.17.

Let us now consider the system presented in Fig. 6.18, which models the original system, working with a simplified system that no longer presents symmetries, but which will have the same eigenfrequencies and the same eigenvectors as the original system.

If eigenfrequencies are calculated for this system, one gets: $2853p$, $2438p$, $2181p$, $1667p$, $1212p$, $780p$, $194p$. The matrix of the eigenmodes is given by (Fig. 6.19):

$$\begin{bmatrix} -0.50 & 0.18 & -0.61 & 0.83 & 0.18 & -0.35 & 0.27 \\ 0.69 & -0.12 & 0.15 & 0.34 & 0.15 & -0.40 & 0.35 \\ -0.19 & -0.21 & -0.02 & -0.20 & -0.20 & -0.54 & 0.37 \\ 0.24 & 0.93 & -0.14 & -0.31 & -0.34 & -0.39 & 0.39 \\ -0.01 & -0.03 & 0.05 & 0.10 & -0.40 & 0.45 & 0.43 \\ -0.41 & 0.21 & 0.72 & 0.11 & 0.56 & -0.04 & 0.39 \\ 0.07 & -0.04 & -0.26 & -0.21 & 0.56 & 0.27 & 0.42 \end{bmatrix}.$$

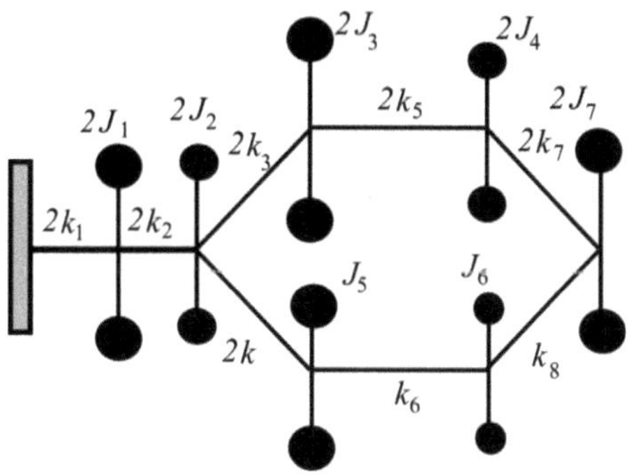

Fig. 6.18 Simplified system

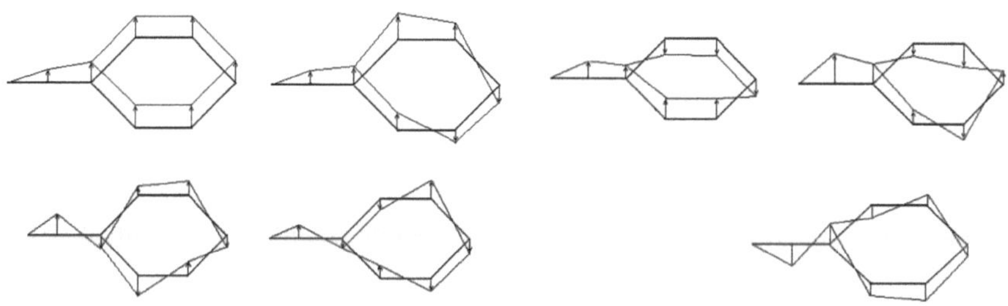

Fig. 6.19 Eigenmodes for the simplified system

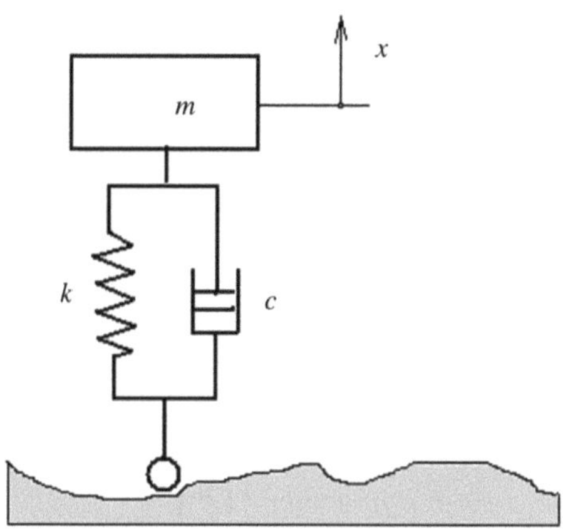

Fig. 6.20 One degree of freedom model

A6 Models for the study of the transverse vibrations of a vehicle.

In the following some simple models to study the vibration of a car are presented (Fig. 6.20).

i. In a first approximation, only the transverse vibrations of the vehicle mass can be considered. The vehicle follows the road profile x_0 with the help of the engine. The motion equation for this simplified model with only one degree of freedom will be:

$$m\ddot{x} + \kappa(x - x_0) = c(\dot{x} - \dot{x}_0) = 0,$$

or

$$m\ddot{x} + c\dot{x} + \kappa x = \underbrace{\kappa x_0 + c\dot{x}_0}_{F(t)}.$$

The system has the eigenfrequency $\omega^2 = \frac{\kappa}{m}$ and the solution of the homogeneous equation:

$$x_0 = Ae^{-\frac{c}{2m}t}\cos(\omega t + \varphi).$$

The solution "declines" quickly over time due to the dampers and the system response can be well approximated by the particular solution x_p obtained using Eq. (6.56).

ii. A better model considering the "pitching" motion of the vehicle is obtained considering the mass sprung on the elastic elements. The motion equations written in the center of the mass will be:

$$m\ddot{x}_C + \kappa_1(x_1 - x_{01}) + \kappa_2(x_2 - x_{02}) + c_1(\dot{x}_1 - \dot{x}_{01}) + c_2(\dot{x}_2 - \dot{x}_{02}) = 0,$$
$$J\ddot{\theta} - \kappa_1 a(x_1 - x_{01}) + \kappa_1 b(x_2 - x_{02}) - c_1 a(\dot{x}_1 - \dot{x}_{01}) + c_2 b(\dot{x}_2 - \dot{x}_{02}) = 0.$$

We have $x_C = \frac{bx_1 + ax_2}{L}, \theta = \frac{x_1 - x_2}{L}$ from where (Fig. 6.21):

$$x_1 \cong x_C - a\theta, x_2 \cong x_C + b\theta.$$

The two equations can be written:

$$\begin{bmatrix} m & 0 \\ 0 & J \end{bmatrix}\begin{Bmatrix} \ddot{x}_C \\ \ddot{\theta} \end{Bmatrix} + \begin{bmatrix} \kappa_1 + \kappa_2 & -\kappa_1 a + \kappa_2 b \\ -\kappa_1 a + \kappa_2 b & \kappa_1 a^2 + \kappa_2 b^2 \end{bmatrix}\begin{Bmatrix} x \\ \theta \end{Bmatrix}$$
$$+ \begin{bmatrix} c_1 + c_2 & -c_1 a + c_2 b \\ -c_1 a + c_2 b & c_1 a^2 + c_2 b^2 \end{bmatrix}\begin{Bmatrix} \dot{x}_C \\ \dot{\theta} \end{Bmatrix} = \begin{Bmatrix} \kappa_1 x_{01} + \kappa_2 x_{02} + c_1\dot{x}_{01} + c_2\dot{x}_{02} \\ -\kappa_1 a x_{01} + \kappa_2 b x_{02} - c_1 a\dot{x}_{01} + c_2 b\dot{x}_{02} \end{Bmatrix}.$$

iii. Coupling and decoupling the coordinates

In the following we will illustrate the importance of coordinate choices in writing motion equations. In order to simplify the presentation we will neglect the damping c.

Fig. 6.21 Two degrees of freedom system

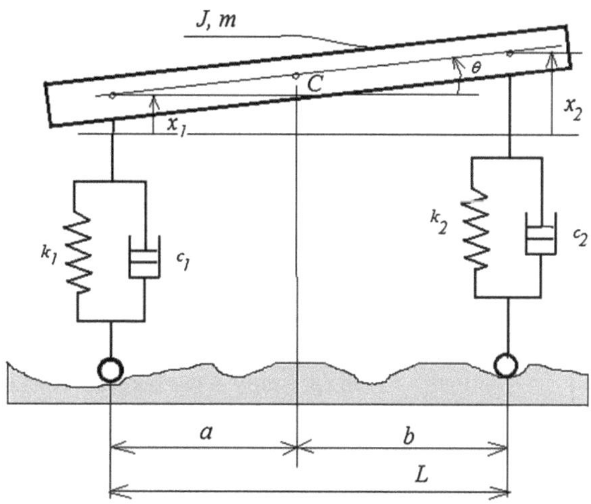

a. For the case ii., if $c_1 = c_2 = 0$ the motion equations result:

$$\begin{bmatrix} m & 0 \\ 0 & J \end{bmatrix} \begin{Bmatrix} \ddot{x} \\ \ddot{\theta} \end{Bmatrix} + \begin{bmatrix} \kappa_1 + \kappa_2 & -\kappa_1 a + \kappa_2 b \\ -\kappa_1 a + \kappa_2 b & \kappa_1 a^2 + \kappa_2 b^2 \end{bmatrix} \begin{Bmatrix} x \\ \theta \end{Bmatrix} = \begin{Bmatrix} \kappa_1 x_{01} + \kappa_2 x_{02} \\ -\kappa_1 a x_{01} + \kappa_2 b x_{02} \end{Bmatrix}.$$

The coupling of the coordinates is made by the non-diagonal elements of the stiffness matrix $-k_1 a + k_2 b \neq 0$. The coupling is called static (or elastic).

b. *A particular case*

If $-k_1 a + k_2 b = 0$ then the system becomes:

$$\begin{bmatrix} m & 0 \\ 0 & J \end{bmatrix} \begin{Bmatrix} \ddot{x} \\ \ddot{\theta} \end{Bmatrix} + \begin{bmatrix} k_1 + k_2 & 0 \\ 0 & k_1 a^2 + k_2 b^2 \end{bmatrix} \begin{Bmatrix} x \\ \theta \end{Bmatrix} = \begin{Bmatrix} k_1 x_{01} + k_2 x_{02} \\ -k_1 a x_{01} + k_2 b x_{02} \end{Bmatrix}.$$

The equations are decoupled both statically and dynamically; the system "breaks" into two independent equations.

c. We choose a point P (at the distance e left to C) so that a vertical force applied at this point produces only a translational motion. Imposing that the resultant moment of forces in P is null, in the static case, one obtains:

$$k_1 x(a - e) = k_2 x(b + e) \Rightarrow k_1 a - k_2 b = (k_1 + k_2)e \text{ and } e = \frac{k_1 a - k_2 b}{k_1 + k_2}.$$

It results:

$$x_1 - x_P = (a - e)\theta; x_2 = x_P + (b + e)\theta.$$

The motion equations written in the center of the mass are:

$$\begin{cases} m(\ddot{x}_P + e\ddot{\theta}) + \kappa_1(x_1 - x_{01}) + \kappa_2(x_2 - x_{02}) = 0 \\ J_C\ddot{\theta} + \kappa_2 b(x_2 - x_{02}) - \kappa_2 a(x_1 - x_{01}) = 0. \end{cases}$$

or:

$$m\ddot{x}_P + me\ddot{\theta} + k_1x_1 + k_2x_2 = k_1x_{01} + k_2x_{02}$$
$$J_C\ddot{\theta} + k_2bx_2 - k_1ax_1 = k_2bx_{02} + k_1ax_{01}.$$

But $J_C = J_P - me^2$ and replacing x_1 and x_2 we obtain:

$$\begin{cases} m\ddot{x}_P + me\ddot{\theta} + (\kappa_1 + \kappa_2)x_P = \kappa_1x_{01} + \kappa_2x_{02}, \\ J_P\ddot{\theta} - me^2\ddot{\theta} + \kappa_2bx_P - \kappa_1ax_P + b(b+e)\kappa_2\theta + a(a-e)\kappa_1\theta = \kappa_2bx_{02} - \kappa_1ax_{01}. \end{cases}$$

Inserting in the second equation the term $me^2\ddot{\theta}$ obtained from the first equation, we have:

$$\begin{cases} m\ddot{x}_P + me\ddot{\theta} + (\kappa_1 + \kappa_2)x_P = \kappa_1x_{01} + \kappa_2x_{02} \\ J_P\ddot{\theta} - me\ddot{x}_P + [\kappa_1(a-e)^2 + \kappa_2(b+e)^2]\theta = \kappa_2bx_{02} - \kappa_1ax_{01} \end{cases},$$

or

$$\begin{bmatrix} m & me \\ me & J_P \end{bmatrix}\begin{Bmatrix} \ddot{x}_P \\ \ddot{\theta} \end{Bmatrix} + \begin{bmatrix} \kappa_1 + \kappa_2 & 0 \\ 0 & \kappa_1(a-e)^2 + \kappa_2(b+e)^2 \end{bmatrix}\begin{Bmatrix} x_P \\ \theta \end{Bmatrix}$$
$$= \begin{Bmatrix} \kappa_1x_{01} + \kappa_2x_{02} \\ \kappa_2bx_{02} - \kappa_1ax_{01} \end{Bmatrix}.$$

The coupling of the coordinates is done through the inertia matrix (dynamic coupling).

d. Static and dynamic coupling.

By choosing the point in which we write the displacement x differently from P and C the equations are coupled both statically and dynamically.

A7 We propose to determine the motion equation for a mass m fixed by identical springs (having the elastic constant $k/2$) by a bar that rotates with the angular velocity ω and the angular acceleration ε (Brain model in a car accident, see Fig. 6.22).

Solution We use d'Alembert's equations:

$$m\{a_a\} = \begin{Bmatrix} F_{ex} \\ F_{ey} \end{Bmatrix},$$

where the absolute acceleration is:

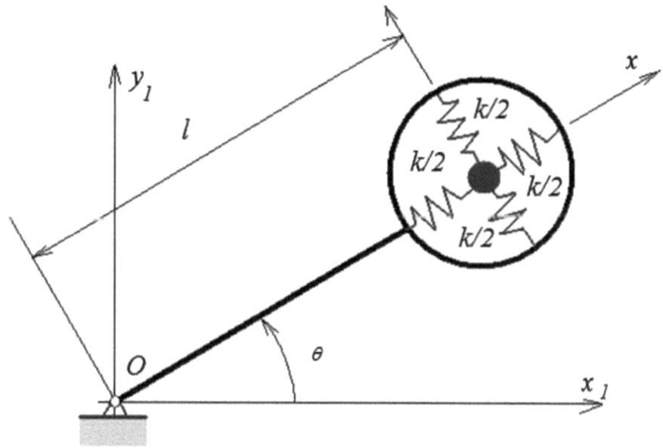

Fig. 6.22 Brain model in a car accident

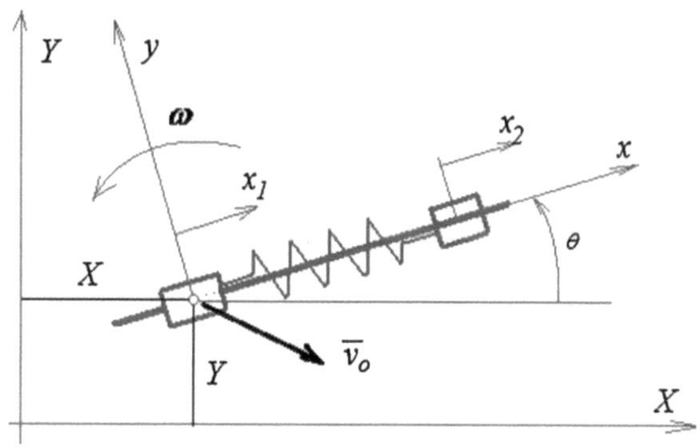

Fig. 6.23 Elastic system with a plane system

$$\{a_a\} = \{a_t\} + \{a_c\} + \{a_r\} = \begin{Bmatrix} \ddot{u} \\ \ddot{v} \end{Bmatrix} + 2\omega \begin{Bmatrix} -\dot{v} \\ \dot{u} \end{Bmatrix} + \varepsilon \begin{Bmatrix} -v \\ l + \dot{u} \end{Bmatrix} - \omega^2 \begin{Bmatrix} l + u \\ v \end{Bmatrix},$$

and the elastic forces:

$$\begin{Bmatrix} F_{ex} \\ F_{ey} \end{Bmatrix} = -k \begin{bmatrix} 1 & 0 \\ 1 & 0 \end{bmatrix} \begin{Bmatrix} u \\ v \end{Bmatrix}.$$

It results:

$$\begin{bmatrix} m & 0 \\ 0 & m \end{bmatrix} \begin{Bmatrix} \ddot{u} \\ \ddot{v} \end{Bmatrix} + 2\omega \begin{bmatrix} 0 & -m \\ m & 0 \end{bmatrix} \begin{Bmatrix} \dot{u} \\ \dot{v} \end{Bmatrix} + \left(\kappa \begin{bmatrix} 1 & 0 \\ 0 & 1 \end{bmatrix} + \varepsilon \begin{bmatrix} 0 & -m \\ m & 0 \end{bmatrix} - \omega^2 \begin{bmatrix} 0 & m \\ m & 0 \end{bmatrix} \right) \begin{Bmatrix} u \\ v \end{Bmatrix}$$
$$= -\varepsilon \begin{bmatrix} 0 & -m \\ m & 0 \end{bmatrix} \begin{Bmatrix} 0 \\ l \end{Bmatrix} + \omega^2 \begin{bmatrix} m & 0 \\ 0 & m \end{bmatrix} \begin{Bmatrix} l \\ 0 \end{Bmatrix}.$$

A8 The bar *AB* executes a plane motion (see Fig. 6.23).

The point O has the coordinates (X, Y). The velocity of this point is $\bar{v}_0 = \dot{X}_0 i + \dot{Y}_0 j$. The angular velocity ω is considered constant, thus $\varepsilon = 0$. Determine the motion equations of the two masses, which are elastically linked with a spring with the rigidity k.

Solution We will use the Lagrange equations in order to obtain the motion equations. To do this, it is necessary to determine the kinetic and the potential energy. The velocities of the two masses are:

$$\{v_1\} = \{v_0\} + [\tilde{\omega}]\{u_1\} + \{\dot{u}_1\} = \begin{Bmatrix} \dot{X}_0 \\ \dot{Y}_0 \end{Bmatrix} + \omega u_1 \begin{Bmatrix} s \\ c \end{Bmatrix} + \dot{u}_1 \begin{Bmatrix} c \\ s \end{Bmatrix}.$$

where $c = \cos\theta$ and $s = \sin\theta$.

$$\{v_2\} = \{v_0\} + [\tilde{\omega}]\{L + u_2\} + \{\dot{u}_2\} = \begin{Bmatrix} \dot{X}_0 \\ \dot{Y}_0 \end{Bmatrix} + \omega \begin{Bmatrix} -(L+u_2)s \\ (L+u_2)c \end{Bmatrix} + \dot{u}_2 \begin{Bmatrix} c \\ s \end{Bmatrix}.$$

We have:

$$\{v_1^2\} = \{v_1\}^T\{v_1\} = (\dot{X}_0 - \omega u_1 s + \dot{u}_1 c)^2 + (\dot{Y}_0 + \omega u_1 c + \dot{u}_1 s)^2,$$
$$\{v_2^2\} = \{v_2\}^T\{v_2\} = \left[\dot{X}_0 - \omega(L+u_2) + \dot{u}_2 c\right]^2 + \left[\dot{Y}_0 + \omega(L+u_2) + \dot{u}_2 s\right]^2,$$

$$\frac{1}{2}\frac{d}{dt}\left[\frac{\partial E_C}{\partial \dot{u}_1}\right] = m_1(\ddot{X}_0 c + \ddot{u}_1 + \ddot{Y}_0 s - \dot{X}_2 s\omega + \dot{Y}_0 c\omega),$$
$$\frac{1}{2}\frac{d}{dt}\left[\frac{\partial E_C}{\partial \dot{u}_2}\right] = m_2(\ddot{X}_0 c + \ddot{u}_2 + \ddot{Y}_0 s - \dot{X}_0 s\omega + \dot{Y}_0 c\omega),$$

$$\frac{\partial E_C}{\partial \dot{u}_1} = m_1(-\dot{X}_0 s\omega + \dot{Y}_0 c\omega + \omega^2 u_1),$$
$$\frac{\partial E_C}{\partial \dot{u}_2} = m_2(-\dot{X}_0 s\omega + \dot{Y}_0 c\omega + \omega^2(L + u_2)),$$

$$E_P = \frac{k(u_1 - u_2)^2}{2}, \; \frac{\partial E_P}{\partial u_1} = k(u_1 - u_2), \; \frac{\partial E_P}{\partial u_2} = k(u_2 - u_1).$$

The equations of motion result as:

$$\begin{bmatrix} m_1 & 0 \\ 0 & m \end{bmatrix}\begin{Bmatrix} \ddot{u}_1 \\ \ddot{u}_2 \end{Bmatrix} + \left(k\begin{bmatrix} 1 & -1 \\ -1 & 1 \end{bmatrix} - \omega^2\begin{bmatrix} m_1 & 0 \\ 0 & m_2 \end{bmatrix}\right)\begin{Bmatrix} u_1 \\ u_2 \end{Bmatrix}$$
$$= -\begin{bmatrix} m_1 & 0 \\ 0 & m_2 \end{bmatrix}\begin{bmatrix} c & s \\ c & s \end{bmatrix}\begin{Bmatrix} \ddot{X}_0 \\ \ddot{Y}_0 \end{Bmatrix} + \omega^2\begin{bmatrix} m_1 & 0 \\ 0 & m_2 \end{bmatrix}\begin{Bmatrix} 0 \\ L \end{Bmatrix}.$$

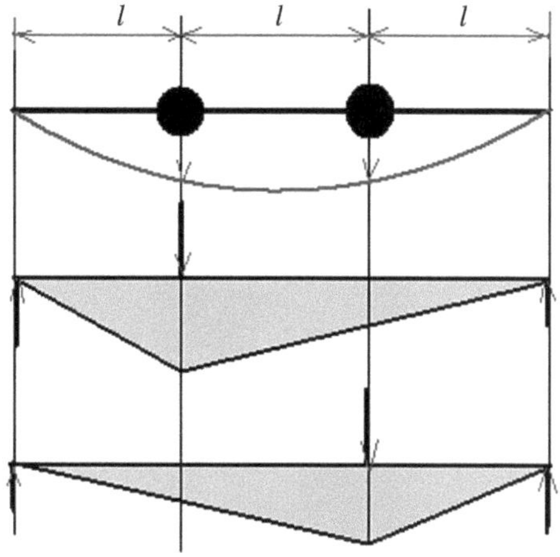

Fig. 6.24 Bending moment diagram

A9 Determine the eigenpulsations of the system presented in Fig. 6.24.

Solution The rigidity matrix is determined using the influence coefficient method:

$$\delta_{11} = \sum \int \frac{M_1^2}{E\,I} dx = \frac{4}{9}\frac{l^3}{E\,I},$$

$$\delta_{12} = \sum \int \frac{M_1 M_2}{E\,I} dx = \frac{7}{18}\frac{l^3}{E\,I},$$

$$\delta_{22} = \sum \int \frac{M_2^2}{E\,I} dx = \frac{4}{9}\frac{l^3}{E\,I},$$

$$[H] = \frac{l^3}{18E\,I}\begin{bmatrix} 8 & 7 \\ 7 & 8 \end{bmatrix},$$

$$[K][H]^{-1} = \frac{18E\,I}{25l^3}\begin{bmatrix} 8 & -7 \\ -7 & 8 \end{bmatrix},$$

$$\{F^{el}\} = -[K]\begin{Bmatrix} y_1 \\ y_2 \end{Bmatrix} = \frac{18E\,I}{25l^3}\begin{bmatrix} 8 & -7 \\ -7 & 8 \end{bmatrix};$$

$$\{F^i\} = -\begin{bmatrix} m & 0 \\ 0 & 2m \end{bmatrix}\begin{Bmatrix} \ddot{y}_1 \\ \ddot{y}_2 \end{Bmatrix},$$

$$\{F^i\} + \{F\} = \{0\}.$$

Thus:

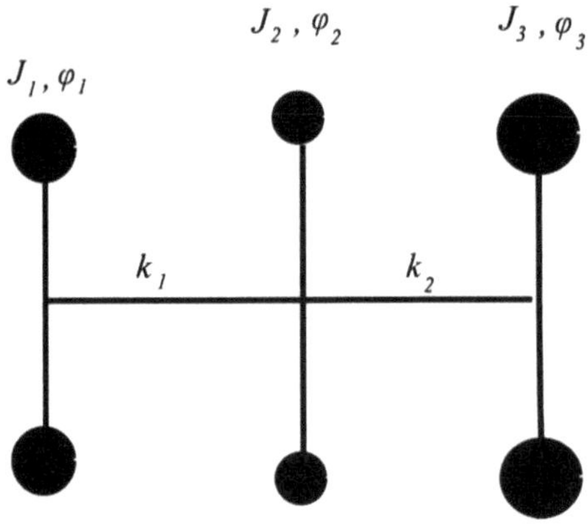

Fig. 6.25 Three-wheel drive transmission

$$m\begin{bmatrix} 1 & 0 \\ 0 & 2 \end{bmatrix}\begin{Bmatrix} \ddot{y}_1 \\ \ddot{y}_2 \end{Bmatrix} + \frac{18}{25}\frac{EI}{l^3}\begin{bmatrix} 8 & -7 \\ -7 & 8 \end{bmatrix}\begin{Bmatrix} y_1 \\ y_2 \end{Bmatrix} = \begin{Bmatrix} 0 \\ 0 \end{Bmatrix}.$$

If we denote:

$$\frac{18}{25}\frac{EI}{ml^3} = \lambda^2,$$

then the characteristic equation is:

$$\begin{vmatrix} 8 - \frac{\omega^2}{\lambda^2} & -7 \\ -7 & 8 - 2\frac{\omega^2}{\lambda^2} \end{vmatrix} = 0,$$

or

$$2\frac{\omega^2}{\lambda^2} - 24\frac{\omega^2}{\lambda^2} + 25 = 0,$$

$$\omega_{1,2}^2 = \frac{12 \pm \sqrt{94}}{2}\lambda^2.$$

A10 Determine the motion equations for a three-wheel drive transmission (see Fig. 6.25):

We will determine the motion equations in two different ways:

(a) Using the d'Alembert equations (see Fig. 6.26):

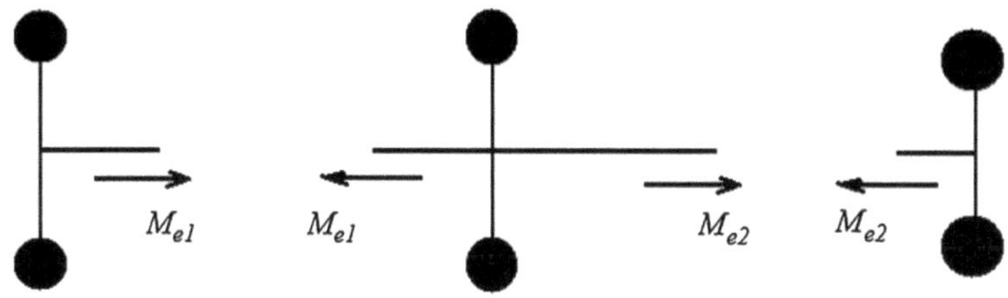

Fig. 6.26 Free-body diagram

$$\begin{cases} -M_1^i - M_{e1} = 0 & M_1^i = J_1\ddot{\varphi}_1, \\ -M_2^i + M_{e1} - M_{e2} = 0 \quad \text{with} & M_2^i = J_2\ddot{\varphi}_2, \\ -M_3^i + M_{e3} = 0 & M_3^i = J_3\ddot{\varphi}_3, \end{cases}$$

$$M_{e1} = k_1(\varphi_1 - \varphi_2),$$
$$M_{e2} = k_2(\varphi_2 - \varphi_3),$$

It results:

$$\begin{cases} J_1\dot{\varphi}_1 + k_1(\varphi_1 - \varphi_2) = 0, \\ J_2\dot{\varphi}_2 - k_1(\varphi_1 - \varphi_2) + k_2(\varphi_2 - \varphi_3) = 0, \\ J_3\dot{\varphi}_3 - k_2(\varphi_2 - \varphi_3) = 0, \end{cases}$$

or

$$\begin{bmatrix} J_1 & & \\ & J_2 & \\ & & J_3 \end{bmatrix} \begin{Bmatrix} \ddot{\varphi}_1 \\ \ddot{\varphi}_2 \\ \ddot{\varphi}_3 \end{Bmatrix} + \begin{bmatrix} k_1 & -k_1 & 0 \\ -k_1 & k_1 + k_2 & -k_2 \\ 0 & -k_2 & k_2 \end{bmatrix} \begin{Bmatrix} \varphi_1 \\ \varphi_2 \\ \varphi_3 \end{Bmatrix} = \begin{Bmatrix} 0 \\ 0 \\ 0 \end{Bmatrix}.$$

(b) The kinetic energy of the system is:

$$E_C = \frac{1}{2}(J_1\dot{\varphi}_1^2 + J_2\dot{\varphi}_2^2 + J_3\dot{\varphi}_3^2) = \begin{bmatrix} \dot{\varphi}_1 & \dot{\varphi}_2 & \dot{\varphi}_3 \end{bmatrix} \begin{bmatrix} J_1 & & \\ & J_2 & \\ & & J_3 \end{bmatrix} \begin{Bmatrix} \dot{\varphi}_1 \\ \dot{\varphi}_2 \\ \dot{\varphi}_3 \end{Bmatrix}.$$

The potential energy is:

$$2E_P = k_1(\varphi_1 - \varphi_2)^2 + k_2(\varphi_2 - \varphi_3)^2$$
$$+ k_2\varphi_1^2 + (k_1 + k_2)\varphi_2^2 + k_2\varphi_3^2 - 2k_1\varphi_1\varphi_2 - 2k_2\varphi_2\varphi_3$$

$$= \left[\varphi_1 \; \varphi_2 \; \varphi_3\right] \begin{bmatrix} k_1 & -k_1 & 0 \\ -k_1 & k_1 + k_2 & -k_2 \\ 0 & -k_2 & k_2 \end{bmatrix} \begin{Bmatrix} \varphi_1 \\ \varphi_2 \\ \varphi_3 \end{Bmatrix}.$$

Using the Lagrange equations we obtain:

$$\frac{d}{dt}\left\{\frac{\partial E_C}{\partial \dot{\varphi}}\right\} = \begin{bmatrix} J_1 & & \\ & J_2 & \\ & & J_3 \end{bmatrix} \begin{Bmatrix} \ddot{\varphi}_1 \\ \ddot{\varphi}_2 \\ \ddot{\varphi}_3 \end{Bmatrix}, \quad \frac{d}{dt}\left\{\frac{\partial E_P}{\partial \dot{\varphi}}\right\} = 0,$$

and

$$\left\{\frac{\partial E_P}{\partial \varphi}\right\} = \begin{bmatrix} k_1 & -k_1 & 0 \\ -k_1 & k_1 + k_2 & -k_2 \\ 0 & -k_2 & k_2 \end{bmatrix} \begin{Bmatrix} \varphi_1 \\ \varphi_2 \\ \varphi_3 \end{Bmatrix}, \quad \left\{\frac{\partial E_C}{\partial \varphi}\right\} = 0.$$

We have used the result $\left\{\frac{\partial U}{\partial X}\right\} = [A]\{X\}$, which is valid if $[A]$ is symmetric and $U = \frac{1}{2}\{X\}^T[A]\{X\}$.

The same motion equations result as in (a).

(c) For this example, it is convenient to change the independent coordinates. The transmission will perform a "rigid" overall motion, and the flywheels will perform relative motion one to each other.

For the rigid motion, it is natural to choose an angle θ_1 so that the moment of momentum of the transmission considered as rigid coincides with the overall moment of momentum of the flywheel. Thus:

$$J_1\dot{\varphi}_1 + J_2\dot{\varphi}_2 + J_3\dot{\varphi}_3 = J\dot{\theta}_1.$$

In this case, we can also take $J_1\varphi_1 + J_2\varphi_2 + J_3\varphi_3 = J\theta_1$ if we consider the angle θ_1, at the initial moment, equal to zero. We can introduce the new coordinates θ_2 and θ_3, which naturally express the relative rotation to the previous wheel, so that:

$$\varphi_1 - \varphi_2 = \theta_2$$
$$\varphi_2 - \varphi_3 = \theta_3.$$

Thus, we have the function change:

$$\begin{Bmatrix} \theta_1 \\ \theta_2 \\ \theta_3 \end{Bmatrix} = \begin{bmatrix} J_1' & J_2' & J_3' \\ 1 & -1 & 0 \\ 0 & 1 & -1 \end{bmatrix} \begin{Bmatrix} \varphi_1 \\ \varphi_2 \\ \varphi_3 \end{Bmatrix},$$

where $J_1' = \frac{J_1}{J}$, $J_2' = \frac{J_2}{J}$, $J_3' = \frac{J_3}{J}$.

We also have the reverse transformation

$$\begin{Bmatrix} \varphi_1 \\ \varphi_2 \\ \varphi_3 \end{Bmatrix} = \begin{bmatrix} 1 & J_2' + J_3' & J_3' \\ 1 & -J_1' & J_3' \\ 0 & -J_1' & -J_1' - J_3' \end{bmatrix} \begin{Bmatrix} \theta_1 \\ \theta_2 \\ \theta_3 \end{Bmatrix} = [L] \begin{Bmatrix} \theta_1 \\ \theta_2 \\ \theta_3 \end{Bmatrix}.$$

We denote:

$$[J] = \begin{bmatrix} J_1 & 0 & 0 \\ 0 & J_2 & 0 \\ 0 & 0 & J_3 \end{bmatrix}.$$

We have:

$$[L]^T [J][L] = \begin{bmatrix} J & 0 & 0 \\ 0 & \frac{J_1(J_2+J_3)}{J} & \frac{J_1 J_3}{I} \\ 0 & \frac{J_1 J_3}{I} & \frac{J_3(J_1+J_2)}{J} \end{bmatrix}$$

and

$$[L]^T [K][L] = \begin{bmatrix} 0 & 0 & 0 \\ 0 & K_2 & 0 \\ 0 & 0 & K_3 \end{bmatrix}.$$

The system becomes expressed in the new coordinates:

$$\begin{bmatrix} J & 0 & 0 \\ 0 & \frac{J_1(J_2+J_3)}{I} & \frac{J_1 J3}{I} \\ 0 & \frac{J_1 J3}{I} & \frac{J_3(J_1+J_2)}{I} \end{bmatrix} \begin{Bmatrix} \ddot{\theta}_1 \\ \ddot{\theta}_2 \\ \ddot{\theta}_3 \end{Bmatrix} + \begin{bmatrix} 0 & 0 & 0 \\ 0 & K_2 & 0 \\ 0 & 0 & K_3 \end{bmatrix} \begin{Bmatrix} \theta_1 \\ \theta_2 \\ \theta_3 \end{Bmatrix} = \{0\},$$

and decouples into:

$$J\ddot{\theta}_1 = 0,$$

which represents the rigid uniform motion $\theta_1 = \omega_1 t + \theta_{10}$ and in the system:

$$\frac{1}{J} \begin{bmatrix} J_1(J_2 + J_3) & J_1 J_3 \\ J_1 J_3 & J_3(J_1 + J_2) \end{bmatrix} + \begin{bmatrix} K_2 & 0 \\ 0 & K_3 \end{bmatrix} \begin{Bmatrix} \theta_1 \\ \theta_2 \end{Bmatrix} = \begin{Bmatrix} 0 \\ 0 \end{Bmatrix}.$$

The system is inertially coupled (dynamic) and elastically uncoupled (static).

A11 Determine the eigenvalues for the system presented in Fig. 6.27.

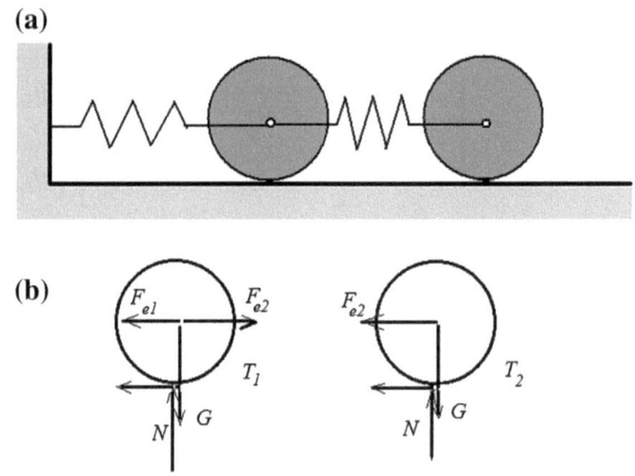

Fig. 6.27 Two rolling cylinders: a) System with two rolling cylinders b)Free-body diagram

Solution The system has four degrees of freedom, each cylinder having the possibility of independent translation and rotation. Using the free-body diagram, it is possible to write the motion equation for every body of the system:

$$\begin{aligned}
&F_{e2} - F_{e1} - T_1 = ma_1, \\
&T_1 R = J\varepsilon_1, \\
&-F_{e2} - T_2 = ma_2, \\
&T_2 R = J\varepsilon_2,
\end{aligned}$$

where $F_{e1} = kx_1$ and $F_{e2} = k(x_2 - x_1)$. Or

$$\begin{bmatrix} m & & & \\ & J & & \\ & & m & \\ & & & J \end{bmatrix} \begin{Bmatrix} a_1 \\ \varepsilon_1 \\ a_2 \\ \varepsilon_2 \end{Bmatrix} = \begin{Bmatrix} F_{e2} - F_{e1} - T_1 \\ T_1 R \\ -F_{e2} - T_2 \\ T_2 R \end{Bmatrix},$$

from where:

$$\begin{bmatrix} m & & & \\ & J & & \\ & & m & \\ & & & J \end{bmatrix} \begin{Bmatrix} a_1 \\ \varepsilon_1 \\ a_2 \\ \varepsilon_2 \end{Bmatrix} + \begin{bmatrix} 2k & -k \\ 0 & 0 \\ -k & k \\ 0 & 0 \end{bmatrix} \begin{Bmatrix} x_1 \\ x_2 \end{Bmatrix} = \begin{Bmatrix} -T_1 \\ T_1 R \\ -T_2 \\ T_2 R \end{Bmatrix}.$$

The kinematical conditions will establish the final form of the equations. We will consider in the following only the case where we have pure rolling. We have the following motion mode:

$$\begin{Bmatrix} v_1 \\ \omega_1 \\ v_2 \\ \omega_2 \end{Bmatrix} = \begin{bmatrix} 1 & 0 \\ \frac{1}{R} & 0 \\ 0 & 1 \\ 0 & \frac{1}{R} \end{bmatrix} \begin{Bmatrix} v_1 \\ v_2 \end{Bmatrix},$$

because there are the following relationships between angular and linear velocities: $\omega_1 = \frac{v_1}{R_1}$ and $\omega_2 = \frac{v_2}{R_2}$.

By differentiation, the relations between accelerations are obtained:

$$\begin{Bmatrix} a_1 \\ \varepsilon_1 \\ a_2 \\ \varepsilon_2 \end{Bmatrix} = \begin{bmatrix} 1 & 0 \\ \frac{1}{R} & 0 \\ 0 & 1 \\ 0 & \frac{1}{R} \end{bmatrix} \begin{Bmatrix} a_1 \\ a_2 \end{Bmatrix}.$$

Using these conditions in the motion equations, we are led to:

$$\begin{bmatrix} m & & & \\ & J & & \\ & & m & \\ & & & J \end{bmatrix} \begin{bmatrix} 1 & 0 \\ \frac{1}{R} & 0 \\ 0 & 1 \\ 0 & \frac{1}{R} \end{bmatrix} \begin{Bmatrix} a_1 \\ a_2 \end{Bmatrix} + \begin{bmatrix} 2k & -k \\ 0 & 0 \\ -k & k \\ 0 & 0 \end{bmatrix} \begin{Bmatrix} x_1 \\ x_2 \end{Bmatrix} = \begin{Bmatrix} -T_1 \\ T_1 R \\ -T_2 \\ T_2 R \end{Bmatrix}.$$

In order to transform the vector of the generalized forces into the displacement space, the system will be multiplied with the matrix:

$$\begin{bmatrix} 1 & \frac{1}{R} & 0 & 0 \\ 0 & 0 & 1 & \frac{1}{R} \end{bmatrix}.$$

It results:

$$\begin{bmatrix} 1 & \frac{1}{R} & 0 & 0 \\ 0 & 0 & 1 & \frac{1}{R} \end{bmatrix} \begin{bmatrix} m & & & \\ & J & & \\ & & m & \\ & & & J \end{bmatrix} \begin{bmatrix} 1 & 0 \\ \frac{1}{R} & 0 \\ 0 & 1 \\ 0 & \frac{1}{R} \end{bmatrix} \begin{Bmatrix} a_1 \\ a_2 \end{Bmatrix}$$

$$+ \begin{bmatrix} 1 & \frac{1}{R} & 0 & 0 \\ 0 & 0 & 1 & \frac{1}{R} \end{bmatrix} \begin{bmatrix} 2k & -k \\ 0 & 0 \\ -k & k \\ 0 & 0 \end{bmatrix} \begin{Bmatrix} x_1 \\ x_2 \end{Bmatrix} = \begin{bmatrix} 1 & \frac{1}{R} & 0 & 0 \\ 0 & 0 & 1 & \frac{1}{R} \end{bmatrix} \begin{Bmatrix} -T_1 \\ T_1 R \\ -T_2 \\ T_2 R \end{Bmatrix},$$

or

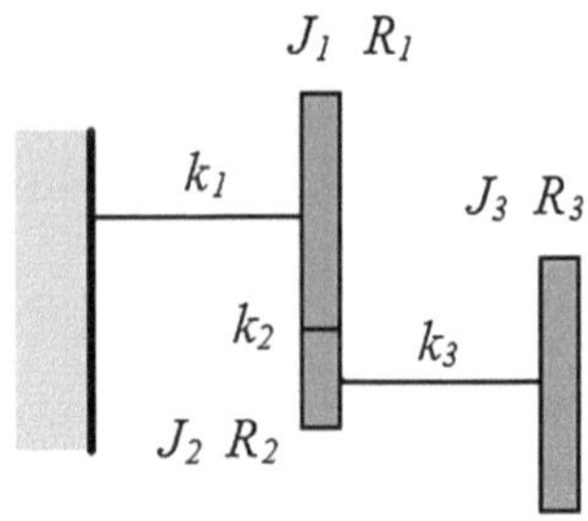

Fig. 6.28 Simple gear model

$$\begin{bmatrix} m+\frac{J}{R^2} & 0 \\ 0 & m+\frac{J}{R^2} \end{bmatrix} \begin{Bmatrix} a_1 \\ a_2 \end{Bmatrix} + \begin{bmatrix} 2k & -k \\ -k & k \end{bmatrix} \begin{Bmatrix} x_1 \\ x_2 \end{Bmatrix} = \begin{Bmatrix} 0 \\ 0 \end{Bmatrix}.$$

The system's eigenvalues are given by the equation:

$$\begin{vmatrix} 2k-\omega^2\left(m+\frac{J}{R^2}\right) & -k \\ -k & k-\omega^2\left(m+\frac{J}{R^2}\right) \end{vmatrix} = 0,$$

$$\omega^4\left(m+\frac{J}{R^2}\right)^2 - 3k\omega^2\left(m+\frac{J}{R^2}\right) + k^2 = 0,$$

with the solution:

$$\omega_{1,2}^2 = \frac{3k\left(m+\frac{J}{R^2}\right) \pm \sqrt{9k^2\left(m+\frac{J}{R^2}\right)^2 - 4\left(m+\frac{J}{R^2}\right)^2 k^2}}{2\left(m+\frac{J}{R^2}\right)^2} = \frac{k(3\pm\sqrt{5})}{2\left(m+\frac{J}{R^2}\right)}.$$

A12 Determine the inertia matrix for the gearing system of Fig. 6.28.

Solution The toothed wheels 1 and 2 form a gear with elastic teeth ($k_2 \neq \infty$). If we denote by $\omega_1, \omega_2, \omega_3$ the angular speeds of wheels 1, 2 and 3, then we obtain the following kinetic energy:

$$E_c = \frac{1}{2}\left[\omega_1 \; \omega_2 \; \omega_3\right] \begin{bmatrix} J_1 & & \\ & J_2 & \\ & & J_3 \end{bmatrix} \begin{Bmatrix} \omega_1 \\ \omega_3 \\ \omega_3 \end{Bmatrix}.$$

We can consider several cases:

(a) If we consider the teeth and the shaft that link the flywheel 2 and 3 as rigid ($k_2 = k_3 = \infty$), the system has one degree of freedom. Since $\omega_1 R_1 = \omega_2 R_2$; $\omega_2 = \omega_3$, thus:

$$\begin{Bmatrix} \omega_1 \\ \omega_3 \\ \omega_3 \end{Bmatrix} = \begin{Bmatrix} 1 \\ \frac{R_1}{R_2} \\ \frac{R_1}{R_2} \end{Bmatrix} \omega_1,$$

$$2E_c = \omega_1^2 \begin{bmatrix} 1 & \frac{R_1}{R_2} & \frac{R_1}{R_2} \end{bmatrix} \begin{bmatrix} J_1 & & \\ & J_2 & \\ & & J_3 \end{bmatrix} \begin{Bmatrix} 1 \\ \frac{R_1}{R_2} \\ \frac{R_1}{R_2} \end{Bmatrix}$$

$$= \omega_1^2 \left[J_1 + (J_2 + J_3)\left(\frac{R_1}{R_2}\right)^2 \right],$$

$$J_{red} = J_1 + (J_2 + J_3)\left(\frac{R_1}{R_2}\right)^2.$$

The matrix of masses becomes, in this case, a scalar and represents the reduced moment of inertia of the system.

(b) We consider now perfectly rigid teeth ($k_2 = \infty$). The system will have two degrees of freedom in this case. We will have only one condition for the speeds: $\omega_1 R_1 = \omega_2 R_2$. It results:

$$\begin{Bmatrix} \omega_1 \\ \omega_3 \\ \omega_3 \end{Bmatrix} = \begin{Bmatrix} 1 & 0 \\ \frac{R_1}{R_2} & 0 \\ 0 & \frac{R_1}{R_2} \end{Bmatrix} \begin{Bmatrix} \omega_1 \\ \omega_3 \end{Bmatrix}.$$

We refer the motion of the flywheel 3 to the motion of the flywheel 1 by the unknown φ_3^* using the relation:

$$R_2\varphi_3 = R_2\varphi_3^*,$$

so

$$\varphi_3^* = \frac{R_2}{R_1}\varphi_3,$$

where φ_3^* represents the angle with which the flywheel 1 would rotate at one rotation φ_3 of the flywheel 3, assuming that the coupling between the wheels 2 and 3 are rigid. We also have:

$$\omega_3^* = \frac{R_2}{R_1}\omega_3,$$

or

$$\begin{Bmatrix} \omega_1 \\ \omega_3 \\ \omega_3 \end{Bmatrix} = \begin{Bmatrix} 1 & 0 \\ \frac{R_1}{R_2} & 0 \\ 0 & \frac{R_1}{R_2} \end{Bmatrix} \begin{Bmatrix} \omega_1 \\ \omega_3^* \end{Bmatrix}.$$

It results:

$$E_c = \frac{1}{2}\begin{bmatrix} \omega_1 & \omega_3^* \end{bmatrix}\begin{bmatrix} 1 & \frac{R_1}{R_2} & 0 \\ 0 & 0 & \frac{R_1}{R_2} \end{bmatrix}\begin{bmatrix} J_1 & & \\ & J_2 & \\ & & J_3 \end{bmatrix}\begin{bmatrix} 1 & 0 \\ \frac{R_1}{R_2} & 0 \\ 0 & \frac{R_1}{R_2} \end{bmatrix}\begin{Bmatrix} \omega_1 \\ \omega_3^* \end{Bmatrix}$$

$$= \begin{bmatrix} \omega_1 & \omega_3^* \end{bmatrix}\begin{bmatrix} J_1 + J_2\left(\frac{R_1}{R_2}\right)^2 & 0 \\ 0 & J_3\left(\frac{R_1}{R_2}\right)^2 \end{bmatrix}\begin{Bmatrix} \omega_1 \\ \omega_3^* \end{Bmatrix}.$$

The inertia matrix is, in this case:

$$[J] = \begin{bmatrix} J_1 + J_2\left(\frac{R_1}{R_2}\right)^2 & 0 \\ 0 & J_3\left(\frac{R_1}{R_2}\right)^2 \end{bmatrix}.$$

(c) If we consider elastic teeth, then the relationship $\omega_1 R_1 = \omega_2 R_2$ is no longer preserved, and ω_1, ω_2 and ω_3 become independent. We refer the rotation of wheel 2 and 3 to the rotation of the flywheel 1 with the relation:

$$R_1\varphi_2^* = R_2\varphi_2 \text{ and } R_1\varphi_3^* = R_2\varphi_3.$$

It results the new independent coordinates:

$$\varphi_2^* = \frac{R_2}{R_1}\varphi_2, \varphi_3^* = \frac{R_2}{R_1}\varphi_3,$$

and thus

$$\omega_2^* = \frac{R_2}{R_1}\omega_2, \omega_3^* = \frac{R_2}{R_1}\omega_3.$$

We have:

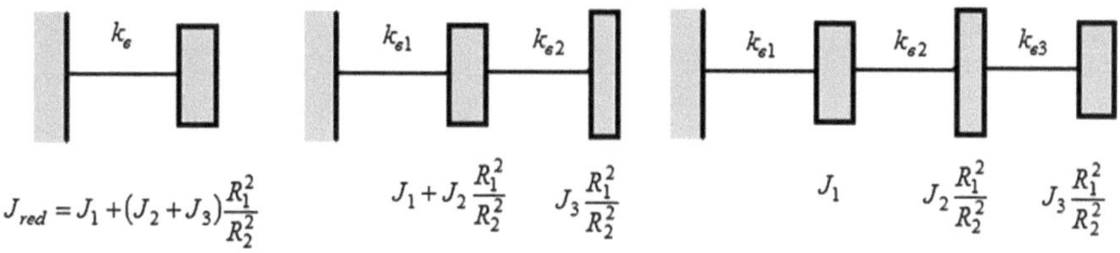

Fig. 6.29 Equivalent mechanical systems of the gear from Fig. 6.28

$$\begin{Bmatrix} \omega_1 \\ \omega_2 \\ \omega_3 \end{Bmatrix} = \begin{Bmatrix} 1 & & 0 \\ & \frac{R_1}{R_2} & \\ 0 & & \frac{R_1}{R_2} \end{Bmatrix} \begin{Bmatrix} \omega_1 \\ \omega_2^* \\ \omega_3^* \end{Bmatrix}.$$

It results

$$[J] = \begin{bmatrix} J_1 & & 0 \\ & J_2\left(\frac{R_1}{R_2}\right)^2 & \\ 0 & & J_3\left(\frac{R_1}{R_2}\right)^2 \end{bmatrix}.$$

The equivalent systems for cases (a), (b), (c) are represented in Fig. 6.28 (Fig. 6.29).

6.4 Vibration of Continuous Bars

6.4.1 Introduction

In the following, a classical presentation of the results known in the discipline will be made. A bar may have transverse vibrations, axial vibrations and torsional vibrations (Ivan 1985; Gillich et al. 2012; Den Hartog 1960; Douglas 2012). The basics of these kinds of vibrations will be presented in the following.

6.4.2 Transverse Vibration of a Bar

Straight, homogeneous, prismatic bars with a symmetry plane will vibrate in this plane (Landau and Lifchitz 1967; Vlase 2012; Vlase and Teodorescu 2013). If the coordinate x is along the bar, y is the axis in which the movement occurs and z is the

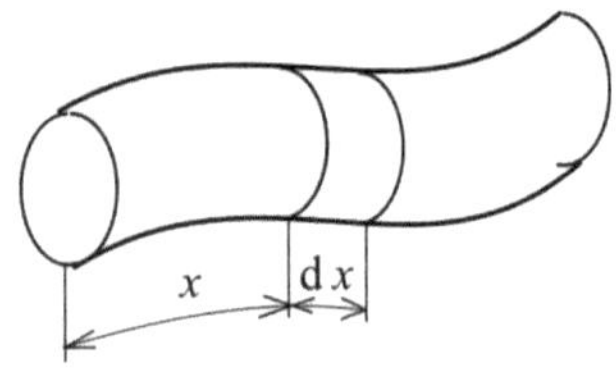

Fig. 6.30 Transverse vibration of a straight bar

neutral axis of the section that we then have in the literature (Buzdugan and Fetcu 1982; Voinea et al. 1983; Meirovitch 1986). The average deformed fiber equation is (Fig. 6.30):

$$\frac{d^2y}{dx^2} = -\frac{M(x)}{EI_z} \tag{6.58}$$

In this relation $M(x)$ is the bending moment in section x, E is the Young's modulus and I_z is the axial geometric moment of inertia around the z-axis. The shear force in the section x is:

$$\frac{dM(x)}{dx} = T(x), \tag{6.59}$$

and the distribution of shear forces along the length of the bar is given by:

$$\frac{dT(x)}{dx} = -p(x). \tag{6.60}$$

Let us now denote by $v(x, t)$ the displacement of the beam to the equilibrium position (the point located on the neutral axis at the distance x from the origin and at time t). The distributed load $p(x)$ can be considered composed by the external load $p_o(x, t)$ and the elementary inertial force $-\rho A \frac{\partial^2 v(x,t)}{\partial t^2}$. By introducing into the equation of the average deformed fiber, we obtain the fourth-order differential equation:

$$\frac{\partial^4 v(x,t)}{\partial x^4} = -\frac{\rho A}{EI_z}\frac{\partial^2 v(x,t)}{\partial t^2} + \frac{p_o(x,t)}{EI_z}. \tag{6.61}$$

In the obtained equation, the effect of the rotation and sliding of the section was neglected. The function $v(x, t)$ must satisfy the boundary conditions. In engineering practice some common boundary conditions, in any section x_o are:

The endpoint is clamped: the displacement and slope must be zero. We have two conditions:

$$v(x_o, t) = 0 \text{ and } \left.\frac{\partial v}{\partial x}\right|_{x=x_0} = 0. \tag{6.62}$$

Simply supported at both ends: the displacement is zero and the bending moment is also zero. The following boundary conditions apply:

$$v(x_o, t) = 0 \text{ and } \left.\frac{\partial^2 v}{\partial x^2}\right|_{x=x_0} = 0. \quad (6.63)$$

Free end: both the shear force and the bending moment are null:

$$\left.\frac{\partial^3 v}{\partial x^3}\right|_{x=x_0} = 0 \text{ and } \left.\frac{\partial^2 v}{\partial x^2}\right|_{x=x_0} = 0. \quad (6.64)$$

Elastically clamped end: The displacement is zero but the slope is proportional to the bending moment:

$$v(x_o, t) = 0 \text{ and } EI_z \left.\frac{\partial^2 v}{\partial x^2}\right|_{x=x_0} = k\frac{\partial v}{\partial x}. \quad (6.65)$$

Elastic support: The displacement is directly proportional to the shear force and the bending moment is zero at the end of the bar:

$$-EI_z \left.\frac{\partial^3 v}{\partial x^3}\right|_{x=x_0} = kv(x_o, t) \text{ and } \frac{\partial^2 v}{\partial x^2}\bigg| x = x_0 = 0. \quad (6.66)$$

6.4.3 Eigenvalues and Eigenmodes in Transverse Vibration

According to the general theory of differential equations with constant coefficients (6.61), a solution of the differential equation is sought, assuming the form (Vlase 2017a, b; Buzdugan et al. 1982; Douglas 2012; Gillich et al. 2012; Ivan 1985):

$$v(x, t) = \Phi(x)\sin(pt + \theta). \quad (6.67)$$

In this relation, $\Phi(x)$ is a function that represents the deformed shape of the bar and p is the circular frequency. If we introduce (6.67) into Eq. (6.61), in the hypothesis $p_o(x, t)$, it is obtained:

$$\frac{d^4\Phi(x)}{dx^4} - \frac{\rho A}{EI_z}p^2\Phi(x) = 0, \quad (6.68)$$

which represents a fourth-order differential equation in x with constant coefficients. Using the notation

$$\lambda^4 = \frac{\rho A}{EI_z}, \quad (6.69)$$

Equation (6.68) becomes:

$$\frac{d^4\Phi(x)}{dx^4} - \lambda^4 p^2 \Phi(x) = 0, \tag{6.68'}$$

and this equation has the solution (see Javanbakht and Öchsner 2017; Öchsner 2016, 2018; Myint-U 1977):

$$\Phi(x) = C_1 \sin(\lambda\sqrt{p}x) + C_2 \cos(\lambda\sqrt{p}x) + C_3 sh(\lambda\sqrt{p}x) + C_4 ch(\lambda\sqrt{p}x). \tag{6.70}$$

The constants C_1, C_2, C_3 and C_4 are obtained from the boundary conditions and there may be, for different sets of boundary conditions (6.62)–(6.66), used for the two ends, fifteen distinct sets of constant values.

The first three derivatives of the function (6.70) are:

$$\Phi'(x) = \lambda\sqrt{p}\left[C_1 \cos(\lambda\sqrt{p}x) - C_2 \sin(\lambda\sqrt{p}x) + C_3 ch(\lambda\sqrt{p}x) + C_4 sh(\lambda\sqrt{p}x)\right], \tag{6.71}$$

$$\Phi''(x) = p\lambda^2\left[-C_1 \sin(\lambda\sqrt{p}x) - C_2 \cos(\lambda\sqrt{p}x) + C_3 sh(\lambda\sqrt{p}x) + C_4 ch(\lambda\sqrt{p}x)\right], \tag{6.72}$$

$$\Phi'''(x) = \lambda^3 p\sqrt{p}\left[-C_1 \cos(\lambda\sqrt{p}x) + C_2 \sin(\lambda\sqrt{p}x) + C_3 ch(\lambda\sqrt{p}x) + C_4 sh(\lambda\sqrt{p}x)\right]. \tag{6.73}$$

Let us illustrate the case of a bar with the length l, clamped at both ends. We must respect, at both ends, the boundary conditions (6.62):

$$v(x_o, t) = 0, \quad \left.\frac{\partial v}{\partial x}\right|_{x=0} = 0,$$

thus

$$v(x_o, t) = 0, \quad v(0, t) = 0, \quad \left.\frac{\partial v}{\partial x}\right|_{x=0} = 0, \quad \left.\frac{\partial v}{\partial x}\right|_{x=l} = 0, \tag{6.74}$$

$$C_2 + C_4 = 0$$

$$C_1 \sin(\lambda\sqrt{pl}) + C_2 \cos(\lambda\sqrt{pl}) + C_3 sh(\lambda\sqrt{pl}) + C_4 ch(\lambda\sqrt{pl}) = 0,$$

$$C_1 + C_3 = 0,$$

$$C_1 \cos(\lambda\sqrt{pl}) - C_2 \sin(\lambda\sqrt{pl}) + C_3 ch(\lambda\sqrt{pl}) + C_4 sh(\lambda\sqrt{pl}) = 0. \tag{6.75}$$

The condition that the homogeneous linear system (6.75) has solutions, others than the null solution, is that:

$$\begin{vmatrix} 0 & 1 & 0 & 1 \\ 1 & 0 & 1 & 0 \\ \sin(\lambda\sqrt{p}l) & \cos(\lambda\sqrt{p}l) & sh(\lambda\sqrt{p}l) & ch(\lambda\sqrt{p}l) \\ \cos(\lambda\sqrt{p}l) & -\sin(\lambda\sqrt{p}l) & ch(\lambda\sqrt{p}l) & sh(\lambda\sqrt{p}l) \end{vmatrix} = 0, \tag{6.76}$$

or

$$ch(\lambda\sqrt{p}l)\cos(\lambda\sqrt{p}l) - 1 = 0. \tag{6.77}$$

The transcendent Eq. (6.77) has an infinite number of roots:

$$\beta_n = \lambda\sqrt{p_n}l \quad \text{where } n = \overline{1, \infty}, \tag{6.78}$$

from where it is possible to obtain the eigenpulsations:

$$p_n = \frac{\beta_n^2}{\lambda^2 l^2} = \frac{\beta_n^2}{l^2}\sqrt{\frac{EI_z}{\rho A}}. \tag{6.79}$$

The eigenfunctions are given by:

$$\Phi_n(x) = C_1 \sin(\lambda\sqrt{p_n x}) + C_2 \cos(\lambda\sqrt{p_n x}) + C_3 sh(\lambda\sqrt{p_n x}) + C_4 ch(\lambda\sqrt{p_n x}).$$

The eigenmodes are given by:

$$v_n(x,t) = \begin{bmatrix} C_1 \sin(\lambda\sqrt{p_n x}) + C_2 \sin(\lambda\sqrt{p_n x}) \\ +C_3 sh(\lambda\sqrt{p_n x}) + C_4 ch(\lambda\sqrt{p_n x}) \end{bmatrix} \sin(pt + \theta), \tag{6.81}$$

where constants C_1, C_2, C_3 and C_4 are determined to an arbitrary factor.

The eigenfunctions are orthogonal to each other, that is, we have:

$$\int_0^l \Phi_i(x)\Phi_j(x)\mathrm{d}x = 0 \text{ for } i \neq j. \tag{6.82}$$

The general solution of the differential Eq. (6.68) is:

$$v(x,t) = \sum_{n=1}^{\infty} \Phi_n(x)\sin(p_n t + \theta_n). \tag{6.83}$$

The initial conditions $v(x,0) = f(x)$ and $\frac{\partial v}{\partial t}\big|_{t=0} = g(x)$ lead to the conditions:

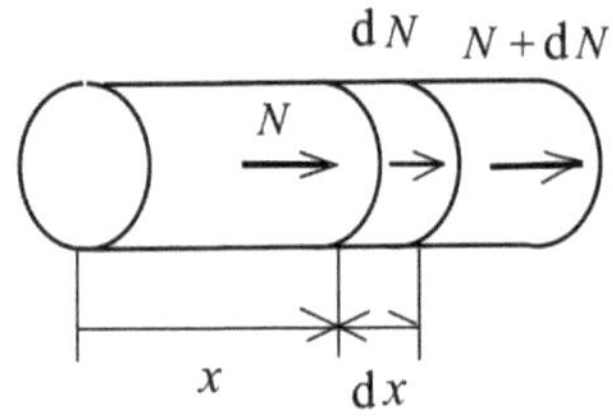

Fig. 6.31 Axial vibration of a straight bar

$$v(x,0)=\sum_{n=1}^{\infty}\Phi_n(x)\sin\theta_n=f(x),$$
$$\left.\frac{\partial v}{\partial t}\right|_{t=0}=\sum_{n=1}^{\infty}p_n\Phi_n(x)\cos\theta_n=g(x), \tag{6.84}$$

which must allow to obtain the constants that appear in the eigenfunctions.

6.4.4 Axial Vibrations of Bars

Let us consider a bar with a constant cross section A (see Fig. 6.31) with length l, Young's modulus E and density ρ. Isolate an infinitesimal element of length $\mathrm{d}x$ at distance x from the left end.

The compression/traction force N appearing in section x is related to the specific deformation $\frac{\partial u}{\partial x}$ by the known relationship:

$$N=EA\frac{\partial u}{\partial x}. \tag{6.85}$$

The variation of the inertia force along the length of the bar is:

$$\frac{\partial F_i}{\partial x}=\rho A\frac{\partial^2 u}{\partial x^2}, \tag{6.86}$$

where ρ is the density of the material.

The variation of the axial force along the length of the bar is:

$$\frac{\partial N}{\partial x}=EA\frac{\partial^2 u}{\partial t^2}. \tag{6.87}$$

In this case, the equilibrium dynamic equation becomes:

$$\frac{\partial^2 u}{\partial x^2}=\frac{\rho}{E}\frac{\partial^2 u}{\partial t^2}, \tag{6.88}$$

if the bar is not loaded with axial forces along the length. The second- order partial differential equation describes the free longitudinal vibration of the considered bar. The boundary conditions of the studied system must be known. The solution will be a function $u(x, t)$. The boundary conditions must specify the displacement u and the speed $\dot{u} = \frac{\partial u}{\partial t}$ at the initial moment $t = 0$ in each section of the bar, x.

$$\varphi(x, 0) = f(x) \text{ and } \left.\frac{\partial \varphi}{\partial t}\right|_{t=0} = g(x). \tag{6.89}$$

The boundary conditions refer in this case to the way in which the ends of the bar are connected with other bodies. In engineering practice, some common boundary conditions are:

(a) If the bar is free at both ends, the axial force at the two ends must be zero, so:

$$N(0, t) = EA \left.\frac{\partial u}{\partial x}\right|_{x=0} = 0 \text{ and } N(l, t) = EA \left.\frac{\partial u}{\partial x}\right|_{x=l} = 0. \tag{6.90}$$

(b) For the clamped bar at both ends, the displacement u at both ends must be zero, thus:

$$u(0, t) = 0 \text{ and } u(l, t) = 0. \tag{6.91}$$

(c) If the bar is clamped at the end $x = 0$ and free at the other end $x = l$, then there will be the boundary conditions:

$$u(0, t) \text{ and } N(l, t) = EA \left.\frac{\partial u}{\partial x}\right|_{x=l} = 0. \tag{6.92}$$

6.4.5 *Eigenvalues and Eigenfunction in Axial Vibration*

The solution of the differential equation describing the motion of the free torsional vibration of the bar implies, according to classical theory, the search for the solutions of the form:

$$u(x, t) = \Theta(x) \sin(pt + \theta). \tag{6.93}$$

Providing that this solution satisfies Eq. (6.88), there is obtained:

$$\frac{\partial^2 \Theta}{\partial x^2} \sin(pt + \theta) = -p^2 \frac{\rho}{E} \Theta \sin(pt + \theta), \tag{6.94}$$

or, after some simplifications:

$$\frac{\partial^2 \Theta}{\partial x^2} + p^2 \frac{\rho}{E} \Theta = 0. \tag{6.95}$$

If we use the abbreviation:

$$\alpha^2 = \frac{p^2 \rho}{E} \text{ or } \alpha = p\sqrt{\frac{\rho}{E}}, \tag{6.96}$$

the classic solution is obtained:

$$\Theta(x) = C_1 \sin \alpha x + C_2 \cos \alpha x, \tag{6.97}$$

and

$$\varphi(x, t) = (C_1 \sin \alpha x + C_2 \cos \alpha x) \sin(pt + \theta). \tag{6.98}$$

The integration constants C_1 and C_2 can be determined for the three fundamental boundary condition cases earlier mentioned.

(a) If the bar is free at both ends, the axial force at the two ends must be zero, thus:

$$N(0, t) = EA \left.\frac{\partial u}{\partial x}\right|_{x=0} = 0 \text{ and } N(l, t) = EA \left.\frac{\partial u}{\partial x}\right|_{x=l} = 0, \tag{6.99}$$

whatever it is t, thus,

$$\begin{aligned} C_1 &= \cos 0 - C_2 \sin 0 = 0, \\ C_1 &= \cos \alpha l - C_2 \sin \alpha l = 0, \end{aligned} \tag{6.100}$$

from where:

$$C_1 = 0 \text{ and } C_2 = \sin \alpha l = 0. \tag{6.101}$$

It results the condition:

$$\sin \alpha l = 0,$$

and thus:

$$\alpha l = n\pi, \tag{6.102}$$

from where:

$$\alpha_n = \frac{n\pi}{l} \quad n = 1, 2, 3, \ldots \tag{6.103}$$

There is an infinite number of solutions that verifies the given equation:

$$u(x,t) = C_{2n} \cos \frac{n\pi}{l} x \sin(p_n t + \theta_n), \tag{6.104}$$

where $n = 1, 2, 3, \ldots$

The integration constants C_{2n} will depend on the initial conditions. From Eq. (6.96) it results the pulsation:

$$p_n = \alpha_n \sqrt{\frac{E}{\rho}} = \frac{n\pi}{l} \frac{E}{\rho}. \tag{6.105}$$

The functions

$$\Theta_n(x) = \cos \frac{n\pi}{l} x, \tag{6.106}$$

will be eigenfunctions for the axial vibration (constant C_2 was chosen equal to 1).

The motion described by Eq. (6.106) is called the eigenmode of vibration of order n for the axial (longitudinal) vibrations. One can directly verify that eigenfunctions satisfy the orthogonality conditions:

$$\int_0^l \Theta_n(x)\Theta_m(x)\mathrm{d}x = 0 \text{ for } m \neq n, \tag{6.107}$$

and

$$\int_0^l \Theta_n(x)\Theta_m(x)\mathrm{d}x = \frac{l}{2} \text{for } m = n. \tag{6.108}$$

The cross sections in which the eigenfunctions are zero are called the nodes.

(b) $u(0,t) = 0$ and $u(l,t) = 0$.

For a clamped bar at both ends, the displacements u at the two ends must be zero, thus: $u(0,t) = 0$ and $u(l,t) = 0$.

The first condition leads to the relationship:

$$u(0,t) = (C_1 \sin 0 + C_2 \cos 0) \sin(pt + \theta) = 0,$$

from where:

$$C_2 = 0,$$

and the second condition leads to the relationship:

$$u(l,t) = (C_1 \sin \alpha l + C_2 \cos \alpha l) \sin(pt + \theta) = 0,$$

hence the condition:

$$\alpha l = n\pi. \tag{6.109}$$

It results:

$$\alpha_n = \frac{n\pi}{l} \quad n = 1, 2, 3, \ldots \tag{6.110}$$

$$p_n = \alpha_n \sqrt{\frac{E}{\rho}} = \frac{n\pi}{l}\sqrt{\frac{E}{\rho}}, \tag{6.111}$$

and the eigenfunctions are, in this case:

$$\Theta_n(x) = \sin\frac{n\pi}{l}x, \tag{6.112}$$

where $C_1 = 1$ was chosen.

(c) If the bar is clamped at the end $x = 0$ and free at the other end $x = l$, then there will be the boundary conditions:

$$u(0, t) = 0 \text{ and } N(l, t) = EA\left.\frac{\partial u}{\partial x}\right|_{x=l} = 0. \tag{6.113}$$

It results

$$u(x, t) = (C_1 \sin 0 + C_2 \cos 0x)\sin(pt + \theta) = 0,$$

from where

$$C_2 = 0, \tag{6.114}$$

and

$$\frac{\partial u(x, t)}{\partial x} = \alpha(C_1 \cos\alpha l - C_2 \sin\alpha l)\sin(pt + \theta) = 0.$$

Thus:

$$\cos\alpha l = 0, \tag{6.115}$$

$$\alpha_n = \frac{1}{l}\left(n\pi - \frac{\pi}{2}\right), \quad n = 1, 2, 3, \ldots\ldots, \tag{6.116}$$

$$p_n = \alpha_n\sqrt{\frac{GI_p}{J}} = \frac{1}{l}\left(n\pi - \frac{\pi}{2}\right)\sqrt{\frac{GI_p}{J}}. \tag{6.117}$$

The eigenfunctions are:

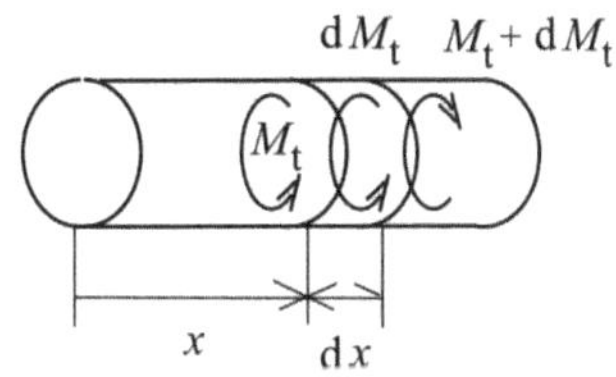

Fig. 6.32 Torsion of a bar

$$\Theta_n(x) = \sin\frac{1}{l}\left(n\pi - \frac{\pi}{2}\right)x. \tag{6.118}$$

6.4.6 Torsional Vibration of the Bar

Let us consider a constant circular cross section bar (see Fig. 6.32), which has the torsional rigidity GI_p and a moment of inertia on the unit of length equal to J. An infinitesimal element of length dx is isolated at the distance x from the end.

Between the torque in the cross section x and the torsion angle φ there is the well known established relationship:

$$M_t = GI_p\frac{\partial\varphi}{\partial x}. \tag{6.119}$$

The moment variation along the bar becomes:

$$\frac{\partial M_t}{\partial x} = GI_p\frac{\partial^2\varphi}{\partial x^2}. \tag{6.120}$$

The variation of the moment of the inertia forces along the bar length is, considering the moment of momentum theorem applied to the infinitesimal element:

$$\frac{\partial M^i}{\partial x} = J_p\frac{\partial^2\varphi}{\partial t^2}. \tag{6.121}$$

If there are no masses attached along the bar length, the moment of inertia J is determined with the relation:

$$J = \rho I_p, \tag{6.122}$$

where ρ is the density of the material.

In this case, the equilibrium dynamic equation becomes:

$$GI_p\frac{\partial^2\varphi}{\partial x^2} = J\frac{\partial^2\varphi}{\partial t^2}, \tag{6.123}$$

if the bar is not loaded with torsional moments along the length. The second-order partial differential equation describes the free torsional vibration of the considered bar. The boundary conditions of the studied system must be known. The solution will be a function $\varphi(x, t)$. The boundary conditions must specify the torsion angle φ and the angular speed $\frac{\partial \varphi}{\partial t}$ at the initial moment $t = 0$ considered in each section of the bar, x.

$$\varphi(x, 0) = f(x) \text{ and } \left.\frac{\partial \varphi}{\partial t}\right|_{t=0} = g(x). \tag{6.124}$$

The boundary conditions refer in this case to the way in which the ends of the bar are linked to the other bodies. There may be three kinds of boundary conditions:

(a) For a free bar at both ends, the torsional moments at both ends must be zero, thus:

$$M_t(0, t) = GI_p \left.\frac{\partial \varphi}{\partial x}\right|_{x=0} = 0 \text{ and } M_t(l, t) = GI_p \left.\frac{\partial \varphi}{\partial x}\right|_{x=l} = 0. \tag{6.125}$$

(b) For the clamped bar at both ends, the slopes of the bar ends must be equal to zero, thus:

$$\varphi(0, t) = 0 \text{ and } \varphi(l, t) = 0. \tag{6.126}$$

(c) If the bar is clamped at the end $x = 0$ and free at the other end $x = l$, then there will be the boundary conditions:

$$\varphi(0, t) = 0 \text{ and } M_t = (l, t) = GI_p \left.\frac{\partial \varphi}{\partial x}\right|_{x=l} = 0. \tag{6.127}$$

6.4.7 *Eigenvalues and Eigenfunctions in Torsional Vibrations*

To solve the differential equation describing the motion of the free torsional vibrations of the bar according to classical theory, solutions are chosen in the form of:

$$\varphi(x, t) = \Psi(x) \sin(pt + \theta). \tag{6.128}$$

Providing that this solution satisfies Eq. (6.123), one gets:

$$GI_p \frac{\partial^2 \Psi}{\partial x^2} \sin(pt + \theta) = -p^2 J \Psi \sin(pt + \theta), \tag{6.129}$$

or, after simplifications:

$$\frac{\partial^2 \Psi}{\partial x^2} + \frac{p^2 J}{G I_p} \Psi = 0. \tag{6.130}$$

If we denote:

$$\alpha^2 = \frac{p^2 J}{G I_p}, \tag{6.131}$$

the classic solution is obtained:

$$\Psi(x) = C_1 \sin \alpha x + C_2 \cos \alpha x, \tag{6.132}$$

and

$$\varphi(x, t) = (C_1 \sin \alpha x + C_2 \cos \alpha x) \sin(pt + \theta). \tag{6.133}$$

The integration constants C_1 and C_2 can be determined for the boundary conditions earlier mentioned.

(a) For the free bar at both ends, torque moments at both ends must be zero, thus:

$$M_t(0, t) = G I_p \frac{\partial \varphi}{\partial x}\bigg|_{x=0} = 0 \text{ and } M_t(l, t) = G I_p \frac{\partial \varphi}{\partial x}\bigg|_{x=l} = 0, \tag{6.134}$$

whatever t is, then

$$\begin{aligned} &C_1 \cos 0 - C_2 \sin 0 = 0, \\ &C_1 \cos \alpha l - C_2 \sin \alpha l = 0, \end{aligned} \tag{6.135}$$

from where:

$$C_1 = 0 \text{ and } C_2 \sin \alpha l = 0 \tag{6.136}$$

It results:

$$\sin \alpha l = 0,$$

from where

$$\alpha l = n\pi, \tag{6.137}$$

so that

$$\alpha_n = \frac{n\pi}{l}, \quad n = 1, 2, 3, \ldots. \tag{6.138}$$

Thus, there is an infinite number of solutions that verifies the given equation, namely

$$\varphi(x,t) = C_{2n} \cos \frac{n\pi}{l} x \sin(p_n t + \theta_n). \tag{6.139}$$

The integration constants C_{2n} will depend on the initial conditions. From the notation for α it results the eigenvalues:

$$p_n = \alpha_n \sqrt{\frac{GI_p}{J}} = \frac{n\pi}{l} \sqrt{\frac{GI_p}{J}}. \tag{6.140}$$

The function

$$\Psi(x) = \cos \frac{n\pi}{l} x, \tag{6.141}$$

is called the eigenfunction (C_2 was chosen equal to 1). The motion described by Eq. (6.141) is called the eigenmode of vibration of the order n. It is verified by direct calculus that the eigenfunctions satisfy the conditions of orthogonality:

$$\int_0^l \Psi_n(x)\Psi_m(x)\mathrm{d}x = 0 \text{ for } m \neq n, \tag{6.142}$$

and

$$\int_0^l \Psi_n(x)\Psi_m(x)\mathrm{d}x = \frac{l}{2} \text{for } m = n. \tag{6.143}$$

Transversal sections in which the eigenfunctions are zero are called the nodes.

(b) For the clamped bar at both ends, the slope at both ends must be equal to zero, thus:

$$\varphi(0,t) = 0 \text{ and } \varphi(l,t) = 0. \tag{6.144}$$

It results:

$$\varphi(0,t)(C_1 \sin 0 + C_2 \cos 0) \sin(pt + \theta) = 0,$$

from where

$$C_2 = 0, \tag{6.145}$$

and

$$\varphi(l,t)(C_1 \sin \alpha l + C_2 \cos \alpha l) \sin(pt+\theta) = 0.$$

That gives us the condition:

$$\alpha l = n\pi. \tag{6.146}$$

As in the previous paragraph:

$$\alpha_n = \frac{n\pi}{l} \quad n = 1, 2, 3, \ldots\ldots \tag{6.147}$$

$$p_n = \alpha_n \sqrt{\frac{GI_p}{J}} = \frac{n\pi}{l}\sqrt{\frac{GI_p}{J}}, \tag{6.148}$$

but their eigenfunctions are in this case:

$$\Psi_n(x) = \sin \frac{n\pi}{l} x, \tag{6.149}$$

where it was chosen: $C_1 = 1$.

(c) If the bar is clamped at the end $x = 0$ and free at the other end $x = l$, then there will be the boundary conditions:

$$\varphi(0,t) = 0 \text{ and } M_t(l,t) = GI_p \left.\frac{\partial \varphi}{\partial x}\right|_{x=l} = 0. \tag{6.150}$$

The first condition gives us:

$$\varphi(x,t) = (C_1 \sin 0 - C_2 \cos 0x) \sin(pt+\theta) = 0.$$

It results:

$$C_2 = 0. \tag{6.151}$$

The second condition gives us:

$$\frac{\partial \varphi(x,t)}{\partial x} = \alpha(C_1 \cos \alpha l - C_2 \sin \alpha l) \sin(pt+\theta) = 0.$$

Thus

$$\cos \alpha l = 0, \tag{6.152}$$

$$\alpha_n = \frac{1}{l}\left(n\pi - \frac{\pi}{2}\right), \quad n - 1, 2, 3, \ldots, \tag{6.153}$$

$$p_n = \alpha_n \sqrt{\frac{GI_p}{J}} = \frac{1}{l}\left(n\pi - \frac{\pi}{2}\right)\sqrt{\frac{GI_p}{J}}. \tag{6.154}$$

The eigenfunction in this case is:

$$\Psi(x) = \sin\frac{1}{l}\left(n\pi - \frac{\pi}{2}\right)x. \tag{6.155}$$

References

Bathe, K.-J., Wilson, E.L.: Numerical Methods in Finite Element Analysis. Prentice-Hall, Englewood Cliffs, New Jersey (1976)

Börm, S., Mehl, C.: Numerical Methods for Eigenvalue Problems. De Gruyter (2012)

Bratu, P.: Vibration of the Elastic Systems, Ed. Tehnica, Bucharest (2000)

Buzdugan, G., Fetcu, L., Rades, M.: Mechanical Vibration. Ed. Did. Ped. Bucharest (1982)

Den Hartog, J.P.: Vibrations mecaniques. Dunod, Paris (1960)

Douglas, T.: Structural Dynamics and Vibrations in Practice: An Engineering Handbook. CRC Press (2012)

Gillich, G.-R., Praisach, Z., Iavornic, C.M.: Reliable method to detect and assess damages in beams based on frequency changes. In: ASME2012 International Design Engineering Technical Conferences and Computers and Information in Engineering Conference, pp. 129–137 (2012)

Harris, C.M., Crede, C.E.: Shock and vibration handbook. Vol.1, 2, 3. McGraw-Hill, New York (1961)

Heylen, W., Lammens, S., Sasa, P.: Modal Analysis. Theory and Testing. Katholieke Universiteit Leuven, Belgium (1997)

Ivan, M.: Bazele calcului liniar al structurilor. Ed. Facla, Timisoara (1985)

Javanbakht, Z., Öchsner, A.: Advanced Finite Element Simulation with MSC Marc: Application of User Subroutines. Springer, Heidelberg (2017)

Landau, L., Lifchitz, E.: Théorie de l´élasticité. Editions Mir, Moscou (1967)

Mangeron, D., Goia, I., Vlase, S.: Symmetrical branched systems vibration, Memoirs of the scientific sections of the romanian academy, Series IV, XII, 1 (1991)

Meirovitch, L.: Elements of Vibration Analysis, 2nd edn. McGraw-Hill, NewYork (1986)

Meirovitch, L.: Principles and Techniques of Vibrations. Pearson (1996)

Myint-U, T.: Ordinary Differential Equations, Elsevier (1977)

Negrean, I.: Advanced notions in analytical dynamics of systems. Acta Technical Napocensis, Series: Applied Mathematics, Mechanics and Engineering, vol. 60, Issue IV, November, p. 491 (2017a)

Negrean, I.: Mass distribution in analytical dynamics of systems. Acta Technical Napocensis, Series: Applied Mathematics, Mechanics and Engineering, vol. 60, Issue II, June, p. 175 (2017b)

Öchsner, A.: Computational Statics and Dynamics—An Introduction Based on the Finite Element Method. Springer, Singapore (2016)

Rades, M.: Identification of The Mechanical Systems. Ed. Academiei (1979)

Rades, M.: Mechanical Vibrations, II. Ed. PRINTECH (2010)

Schmitz, T.L., Scott Smith, K.: Mechanical Vibrations: Modeling and Measurement, Springer, Heidelberg (2012)

Singer, S.F.: Symmetry in Mechanics. Springer, Heidelberg (2004)

Timoshenko, P.S., Gere, J.M.: Theory of Elastic Stability, 2nd edn. McGraw-Hill, NewYork, London (2009)

Timoshenko, S., Zoung, D.H.: Vibration Problems in Engineering. Van Nastrand Company (1955)

Tofan, M., Vlase, S.: Vibration of the Mechanical Systems. Transylvania University Press (1985)

Vlase, S.: A method of eliminating Lagrangean multipliers from the equation of motion of interconnected mechanical system. ASME Trans. J. Appl. Mech. **54**, 235–236 (1987a)

Vlase, S.: Elimination of Lagrangean multipliers. Mech. Res. Commun. **14**, 17–20 (1987b)

Vlase, S.: Dynamical response of a multibody system with flexible elements with a general three-dimensional motion. Rom. J. Phys. **57**(3–4), 676–693 (2012)

Vlase, S., Marin, M., Öchsner, A.: Considerations of the transverse vibration of a mechanical system with two identical bars. In: Proceedings of the Institution of Mechanical Engineers. PartL: Journals of Materials: Design and Applications, November (2017a)

Vlase, S., Marin, M., Scutaru, M.L., Munteanu, R.: Coupled Transverse and Torsional Vibrations in a Mechanical System with Two Identical Beams. AIPAdvances7, 065301 (2017b)

Vlase, S., Teodorescu, P.P.: Elasto-dynamics of a solid with a general "rigid" motion using FEM model. Part I. Theoretical approach. Rom. J. Phys. **58**(7–8), 872–881 (2013)

Voinea, R., Voiculescu, D., Ceausu, V.: Mechanics. Edit. Did. Ped., Bucharest (1983)

Wittenburg, J.: Dynamics of Multibody Systems, 2nd edn. Springer, Heidelberg, New York (2008)

Zingoni, A.: On the symmetries and vibration modes of layered spacegrids. Eng. Struct. **27**, 629–638 (2005)

Chapter 7
Dynamical Systems

7.1 Introduction

This chapter aims to present only some aspects of the dynamical systems related to the subject of this book.

Dynamical systems have been long understood as classical mechanical systems. At present, this notion has a much broader meaning, including electromagnetic, biological, social, political etc. systems (Guckenheimer and Holmes 1983). Generally, the evolution of such a system is characterized by differential equations and the qualitative solution or analysis of these equations is the main results in the field (see Alligood et al. 2000; Anosov and Arnold 1988; Arnold 1973, 2013).

Dynamical systems can be categorized into two classes:

- Continuous dynamical systems, i.e. the mathematical model leads to a differential equation (or a differential equation system) or a partial differential equation (or a system of partial differential equations);
- Discrete dynamical systems, i.e. the mathematical model leads to a recurrence relationship or a system of recurrence equations.

Continuous dynamical systems use first-order differential equation systems to describe them. If a differential equation is of a higher order, then it is equivalent to a system of equations of the first order. Let us consider the differential equation of order n:

$$x^{(n)} + f(x, \dot{x}, \ddot{x}, \ldots, x^{(n-1)}, t) = 0. \tag{7.1}$$

Now let us change the functions as follows:

S. Vlase et al., *Eigenvalue and Eigenvector Problems in Applied Mechanics*, Advanced Structured Materials 96, https://doi.org/10.1007/978-3-030-00991-5_7

$$\begin{aligned}
\dot{x}_1 &= x_2, \\
\dot{x}_2 &= x_3, \\
&\dots\dots\dots\dots\dots\dots \\
\dot{x}_{n-1} &= x_n, \\
\dot{x}_n &= -f(x_1, x_2, \dots, x_{n-1}, x_n, t),
\end{aligned} \tag{7.2}$$

or, in a matrix form:

$$\{\dot{x}\} = \{f(x, t)\}. \tag{7.3}$$

A very much used form in engineering practice is the linear one:

$$\{\dot{x}\} = [A]\{x\}, \tag{7.4}$$

where $[A]$ is a matrix with constant coefficients, which we will analyse in the following.

It is called the phase space, the space R^n of the variables $x_1, x_2, \dots, x_{n-1}, x_n$, where each possible state is represented by a unique point in the state phase.

The set of points defined by the position vectors $\{x(t)\} \in R^n$ is called the phase space trajectory. It is called the graph of the solution of a system of differential equations the set of points:

$$[\{x(t)\}, t] \in R^{n+1}. \tag{7.5}$$

It is called the phase path of a system of differential equations, passing through the point $\{x_0\}$, the totality of the end points of the vectors $\{x(t)\}$ in the phase space, which are solutions of the system of differential equations that for the time $t = t_0$ passes through $\{x_0\}$, thus:

$$\{x(t_0)\} = \{x_0\}. \tag{7.6}$$

7.2 Linear Systems with Two Degrees of Freedom

If the dimension of the system of differential equations with constant coefficients is two, it can be written as:

$$\begin{aligned}
\dot{x} &= a_{11}x + a_{12}y, \\
\dot{y} &= a_{21}x + a_{22}y.
\end{aligned} \tag{7.7}$$

The characteristic equation is:

$$\begin{vmatrix} a_{11}-\lambda & a_{12} \\ a_{21} & a_{22}-\lambda \end{vmatrix} = 0, \tag{7.8}$$

or

$$\lambda^2 - \lambda(a_{11}+a_{22}) + a_{11}a_{22} - a_{12}a_{21} = 0. \tag{7.9}$$

If the roots of the characteristic equation are λ_1 and λ_2, the following cases can be distinguished:

(a) Roots λ_1 and λ_2 are real, positive and distinct. In modal variables ξ_1 and ξ_2, the decoupled differential equations are:

$$\dot{\xi}_1 = \lambda_1 \xi_1, \quad \dot{\xi}_2 = \lambda_2 \xi_2, \tag{7.10}$$

and have the solutions:

$$\xi_1 = \xi_{10} e^{\lambda_1 t} \text{ and } \xi_2 = \xi_{20} e^{\lambda_2 t}. \tag{7.11}$$

The linear system:

$$\begin{bmatrix} a_{11}-\lambda_i & a_{12} \\ a_{21} & a_{22}-\lambda_i \end{bmatrix} \begin{Bmatrix} \Phi_{i1} \\ \Phi_{i2} \end{Bmatrix} = \begin{Bmatrix} 0 \\ 0 \end{Bmatrix}, \quad i = 1, 2 \tag{7.12}$$

offers the eigenvectors (we chose $\Phi_{11} = \Phi_{12} = 1$):

$$\begin{Bmatrix} \Phi_{11} \\ \Phi_{12} \end{Bmatrix} = \begin{Bmatrix} 1 \\ \dfrac{a_{11}-\lambda_1}{a_{12}} \end{Bmatrix}, \begin{Bmatrix} \Phi_{21} \\ \Phi_{22} \end{Bmatrix} = \begin{Bmatrix} 1 \\ \dfrac{a_{11}-\lambda_2}{a_{12}} \end{Bmatrix}. \tag{7.13}$$

The solutions of the initial equation are:

$$\begin{Bmatrix} x \\ y \end{Bmatrix} = \begin{bmatrix} \Phi_{11} & \Phi_{12} \\ \Phi_{21} & \Phi_{22} \end{bmatrix} \begin{Bmatrix} \xi_1 \\ \xi_2 \end{Bmatrix} = \begin{bmatrix} \Phi_{11} & \Phi_{12} \\ \Phi_{21} & \Phi_{22} \end{bmatrix} \begin{Bmatrix} \xi_{10} e^{\lambda_1 t} \\ \xi_{20} e^{\lambda_2 t} \end{Bmatrix} \begin{Bmatrix} \Phi_{11}\xi_{10} e^{\lambda_1 t} + \Phi_{12}\xi_{20} e^{\lambda_2 t} \\ \Phi_{21}\xi_{10} e^{\lambda_1 t} + \Phi_{22}\xi_{20} e^{\lambda_2 t} \end{Bmatrix}, \tag{7.14}$$

and since λ_1 and λ_2 are positive, it is clear that solutions x and y increase over time to infinity. The integration constants are obtained from the initial conditions, that at the initial moment $t = 0$ the displacements are x_0 and y_0:

$$\begin{Bmatrix} x_0 \\ y_0 \end{Bmatrix} = \begin{bmatrix} \Phi_{11} & \Phi_{12} \\ \Phi_{21} & \Phi_{22} \end{bmatrix} \begin{Bmatrix} \xi_{10} \\ \xi_{20} \end{Bmatrix}, \tag{7.15}$$

from where:

$$\begin{Bmatrix} \xi_{10} \\ \xi_{20} \end{Bmatrix} = \begin{bmatrix} \Phi_{11} & \Phi_{12} \\ \Phi_{21} & \Phi_{22} \end{bmatrix}^{-1} \begin{Bmatrix} x_0 \\ y_0 \end{Bmatrix} = \frac{1}{\Phi_{11}\Phi_{22} - \Phi_{12}\Phi_{21}} \begin{bmatrix} \Phi_{22} & -\Phi_{12} \\ -\Phi_{21} & \Phi_{11} \end{bmatrix} \begin{Bmatrix} x_0 \\ y_0 \end{Bmatrix}. \tag{7.16}$$

Any point in space corresponds to a solution. To represent in the phase space this solution should eliminate the time in the equations:

$$\begin{Bmatrix} x \\ y \end{Bmatrix} = \begin{bmatrix} \Phi_{11} & \Phi_{12} \\ \Phi_{21} & \Phi_{22} \end{bmatrix} \begin{Bmatrix} \xi_{10} e^{\lambda_1 t} \\ \xi_{20} e^{\lambda_2 t} \end{Bmatrix}. \tag{7.17}$$

It results:

$$\begin{Bmatrix} e^{\lambda_1 t} \\ e^{\lambda_2 t} \end{Bmatrix} = \begin{bmatrix} \Phi_{11} & \Phi_{12} \\ \Phi_{21} & \Phi_{22} \end{bmatrix}^{-1} \begin{Bmatrix} x \\ y \end{Bmatrix} = \frac{1}{\Delta} \begin{bmatrix} \frac{\Phi_{22}}{\xi_{10}} & -\frac{\Phi_{12}}{\xi_{10}} \\ -\frac{\Phi_{21}}{\xi_{20}} & \frac{\Phi_{11}}{{}^{20}\xi_{20}} \end{bmatrix} \begin{Bmatrix} x \\ y \end{Bmatrix}, \tag{7.18}$$

where: $\Delta = \Phi_{11}\Phi_{22} - \Phi_{12}\Phi_{21}$.

One obtains:

$$\lambda_1 t = \ln \frac{\xi_{10}}{\Delta}(\Phi_{22}x - \Phi_{12}y),\ \lambda_2 t = \ln \frac{\xi_{20}}{\Delta}(-\Phi_{21}x - \Phi_{11}y), \tag{7.19}$$

from where one obtains:

$$\left[\frac{\xi_{10}}{\Delta}(\Phi_{22}x - \Phi_{12}y)\right]^{\lambda_2} = \left[\frac{\xi_{20}}{\Delta}(-\Phi_{21}x + \Phi_{11}y)\right]^{\lambda_1}. \tag{7.20}$$

The representation of these curves is shown in Fig. 7.1 for $\lambda_1 < \lambda_2$. If $\lambda_2 < \lambda_1$ the representation is made in Fig. 7.2. Such a critical point (equilibrium point) is an unstable node.

To get an image consider the particular case:

$$\begin{Bmatrix} \dot{x} \\ \dot{y} \end{Bmatrix} = \begin{bmatrix} \frac{7}{8} & \frac{3}{8} \\ -\frac{3}{8} & \frac{17}{8} \end{bmatrix} \begin{Bmatrix} x \\ y \end{Bmatrix}.$$

The characteristic equation

$$\lambda^2 - 3\lambda + 2 = 0$$

has the solutions $\lambda_1 = 1$ and $\lambda_2 = 2$. The eigenvectors are obtained by solving the linear systems:

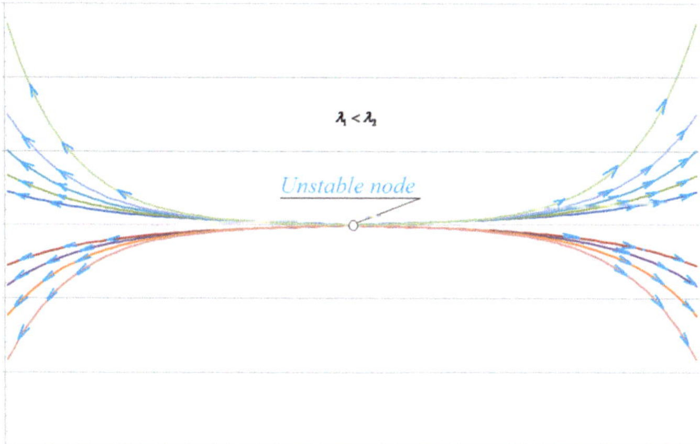

Fig. 7.1 Real, distinct, positive roots. Unstable node ($\lambda_1 < \lambda_2$)

$$\begin{bmatrix} \frac{7}{8} - \lambda_i & \frac{3}{8} \\ -\frac{3}{8} & \frac{17}{8} - \lambda_i \end{bmatrix} \begin{Bmatrix} e_{i1} \\ e_{i2} \end{Bmatrix} = \begin{Bmatrix} 0 \\ 0 \end{Bmatrix}.$$

One obtains:

$$\{e_1\} = \begin{Bmatrix} 3 \\ 1 \end{Bmatrix}; \{e_2\} = \begin{Bmatrix} 1 \\ 3 \end{Bmatrix},$$

and it results the modal matrix: $[\phi] = \begin{bmatrix} 3 & 1 \\ 1 & 3 \end{bmatrix}$.

In modal coordinates the solutions are:

$$\xi_1 = \xi_{10} e^t \quad \text{and} \quad \xi_2 = \xi_{20} e^{2t}.$$

By eliminating t in the two equations, one obtains:

$$\xi_1^2 = \frac{\xi_{10}^2}{\xi_{20}} \xi_2.$$

The initial system solution is:

$$\begin{Bmatrix} x \\ y \end{Bmatrix} = \begin{bmatrix} 3 & 1 \\ 1 & 3 \end{bmatrix} \begin{Bmatrix} \xi_1 \\ \xi_2 \end{Bmatrix} = \begin{bmatrix} 3 & 1 \\ 1 & 3 \end{bmatrix} \begin{Bmatrix} \xi_{10} e^t \\ \xi_{20} e^{2t} \end{Bmatrix} = \begin{Bmatrix} 3\xi_{10} e^t + \xi_{20} e^{2t} \\ \xi_{10} e^t + 3\xi_{20} e^{2t} \end{Bmatrix}.$$

The boundary conditions at the initial moment $t = 0$ are:

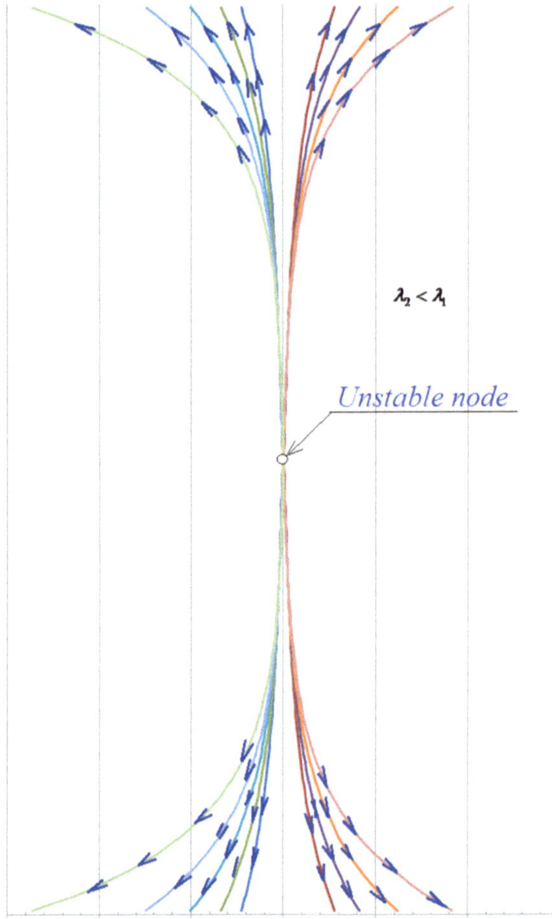

Fig. 7.2 Real, distinct, positive roots. Unstable node ($\lambda_1 < \lambda_2$)

$$x_0 = 3\xi_{10} + \xi_{20}$$
$$y_0 = \xi_{10} + 3\xi_{20},$$

from where:

$$\xi_{10} = \frac{3x_0 - y_0}{8}, \xi_{20} = \frac{-x_0 + 3y_0}{8}.$$

By eliminating t in the two equations, we obtain:

$$(3x - y)^2 = \frac{(3x_0 - y_0)^2}{-x_0 + 3y_0}(-x + 3y).$$

Let us consider the initial conditions: at $t = 0$ we have $x = 1$ and $y = 2$. These conditions are written:

$$1 = 3\xi_{10} + \xi_{20}$$
$$2 = \xi_{10} + 3\xi_{20},$$

from where we obtain the following integration constants:

$$\xi_{10} = \frac{3}{24}, \quad \xi_{20} = \frac{5}{8}.$$

In modal coordinates, solutions become:

$$\xi_1 = \frac{3}{24}e^t, \quad \xi_2 = \frac{5}{8}e^{2t}.$$

and the primary system solution:

$$\begin{Bmatrix} x \\ y \end{Bmatrix} = \begin{bmatrix} 3 & 1 \\ 1 & 3 \end{bmatrix} \begin{Bmatrix} \xi_1 \\ \xi_2 \end{Bmatrix} = \begin{bmatrix} 3 & 1 \\ 1 & 3 \end{bmatrix} \begin{Bmatrix} \xi_{10}e^t \\ \xi_{20}e^{2t} \end{Bmatrix} = \begin{Bmatrix} \frac{9}{24}e^t + \frac{5}{8}e^{2t} \\ \frac{3}{24}e^t + \frac{15}{8}e^{2t} \end{Bmatrix}.$$

Eliminating t in these two equations, one obtains in modal coordinates

$$\xi_1^2 = \frac{1}{40}\xi_2,$$

or in Cartesian coordinates:

$$5(3x - y)^2 = -x + 3y.$$

(b) The roots λ_1 and λ_2 are real, negative and distinct. Things happen at the previous point and the solutions are the same. Only the direction of the curves in the phase space differs.

The representation of these curves is shown in Fig. 7.3 for $|\lambda_1| < |\lambda_2|$. If $|\lambda_2| < |\lambda_1|$ the representation is shown in Fig. 7.4. Such a critical point (equilibrium point) is a stable node. The solutions of the system of equations are:

$$\begin{Bmatrix} x \\ y \end{Bmatrix} = \begin{bmatrix} \Phi_{11} & \Phi_{12} \\ \Phi_{21} & \Phi_{22} \end{bmatrix} \begin{Bmatrix} \xi_1 \\ \xi_2 \end{Bmatrix} = \begin{bmatrix} \Phi_{11} & \Phi_{12} \\ \Phi_{21} & \Phi_{22} \end{bmatrix} \begin{Bmatrix} \xi_{10}e^{\lambda_1 t} \\ \xi_{20}e^{\lambda_2 t} \end{Bmatrix} = \begin{Bmatrix} \Phi_{11}\xi_{10}e^{\lambda_1 t} + \Phi_{12}\xi_{20}e^{\lambda_2 t} \\ \Phi_{21}\xi_{10}e^{\lambda_1 t} + \Phi_{22}\xi_{20}e^{\lambda_2 t} \end{Bmatrix} \tag{7.21}$$

The exponential function will tend to zero if t grows because the two roots λ_1 and λ_2 are negative. So the x and y functions will tend to zero over time. For this reason, we say that the point (0, 0) is a stable node.

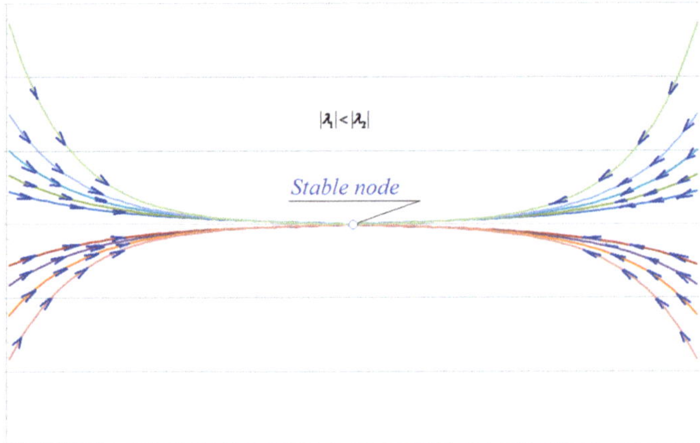

Fig. 7.3 Real, negative, distincte roots. Stable node ($|\lambda|_1 < |\lambda_2|$)

(c) The roots λ_1 and λ_2 are real and with opposite sign. The two solutions will be the same as in the previous case, but while a solution tends to zero, the other tends to infinity. The trajectories of this case are represented in Fig. 7.5.

Axis $O\xi_2$ is called the stable variety because if the initial position is on this axis, the motion is stable. Any other point belongs to an area in which the motion is unstable. If $\lambda_1 < 0$ and $\lambda_2 > 0$ the trajectories have the same representation as in Fig. 7.5 but the orientation changes. In this case, the $O\xi_1$-axis becomes stable. Such a point is called the critical point (equilibrium point) or saddle point and is unstable.

(a) The roots λ_1 and λ_2 are complex conjugated. Then, instead of solutions $e^{(a+bi)t}$ and $e^{(a-bi)t}$, their combinations $\xi_1 = Ae^{at}\cos bt$ and $\xi_1 = Ae^{at}\sin bt$ can be used. The trajectories are spirals, which depart from the origin if $a > 0$, or go to the origin if $a < 0$.

In the first case the critical point (equilibrium point) is an unstable focal point (Fig. 7.6a) and in the second case a stable focal (Fig. 7.6b).

(b) The roots λ_1 and λ_2 are purely imaginary, that is, of form e^{ibt} and e^{-ibt} ($b \neq 0$). The spirals turn into ellipses (Fig. 7.7). The critical (equilibrium) point is simply stable, because the trajectories remain in the vicinity of the node if the time tends to infinity. Such a critical point is called the center.

(c) The roots λ_1 and λ_2 are real and equal. Two situations can be distinguished:

- Case I. The modal matrix is a Jordan block of the following form:

$$\begin{bmatrix} \lambda & 1 \\ 0 & \lambda \end{bmatrix}. \tag{7.22}$$

The system of modal differential equations becomes:

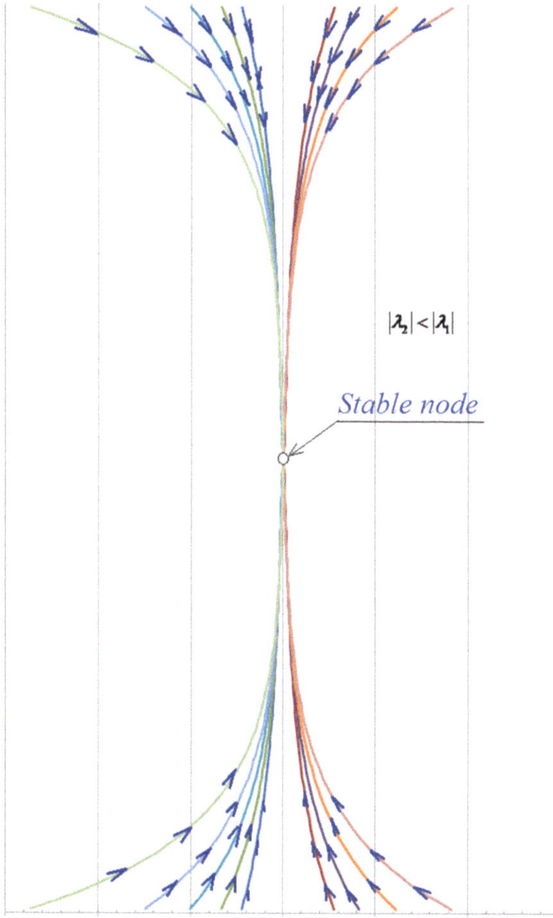

Fig. 7.4 Real, negative, distinct roots. Stable node ($|\lambda|_1 > |\lambda|_2$)

$$\dot{\xi}_1 = \lambda\xi_1 + \xi_2, \quad \dot{\xi}_2 = \lambda\xi_2. \tag{7.23}$$

It results after integration:

$$\xi_2 = \xi_{20}e^{\lambda t} \quad \text{and} \quad \xi_1 = (\xi_{10} + \xi_{20}t)e^{\lambda t}. \tag{7.24}$$

If $\lambda > 0$ then the trajectories are represented in Fig. 7.8a and we are dealing with an unstable node. If $\lambda < 0$ we have a stable node and the trajectories are represented in Fig. 7.8b.

– Case II. The modal matrix is composed of two first-order Jordan blocks.

$$\begin{bmatrix} \lambda & 1 \\ 0 & \lambda \end{bmatrix}. \tag{7.25}$$

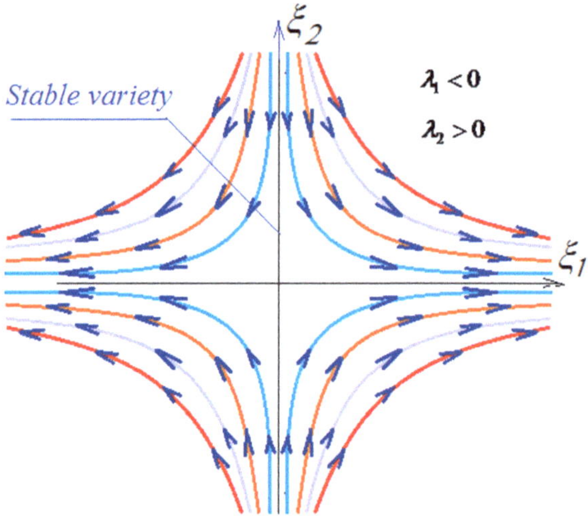

Fig. 7.5 Real, with different signs, roots

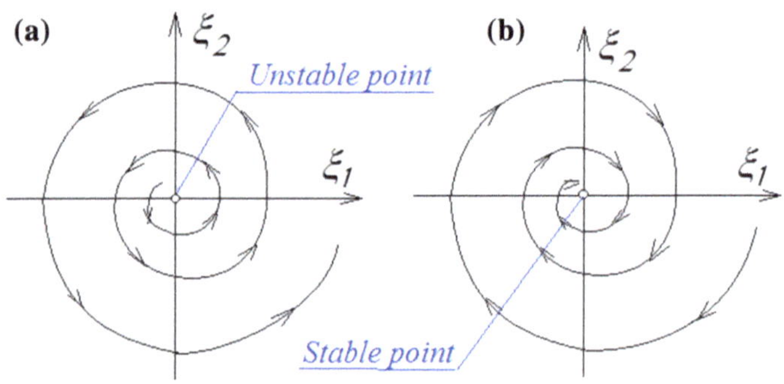

Fig. 7.6 Complex conjugated roots

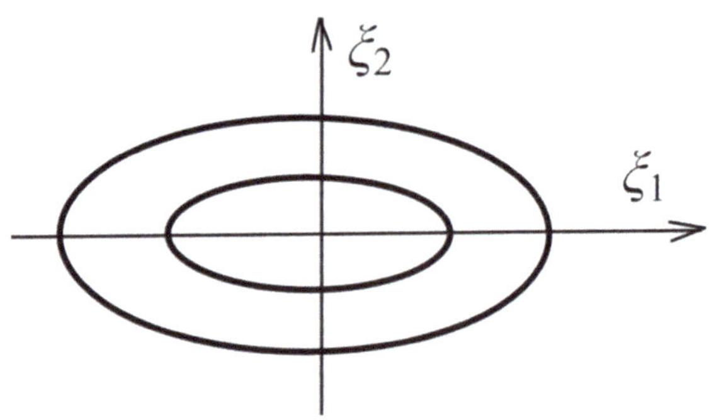

Fig. 7.7 Purely complex roots

The system of modal differential equations becomes:

$$\dot{\xi}_1 = \lambda \xi_1, \quad \dot{\xi}_2 = \lambda \xi_2. \tag{7.26}$$

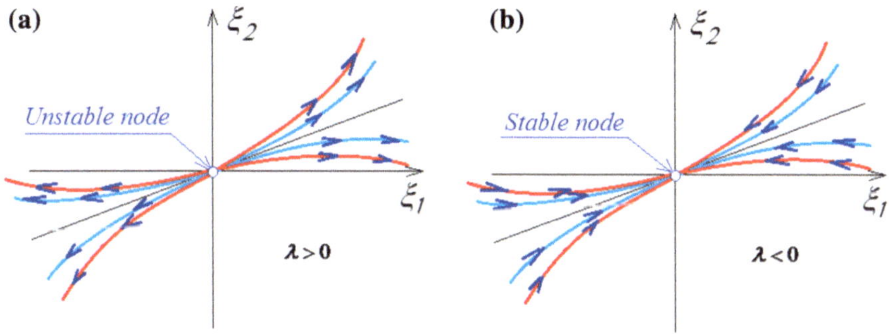

Fig. 7.8 Real, equal roots, case I

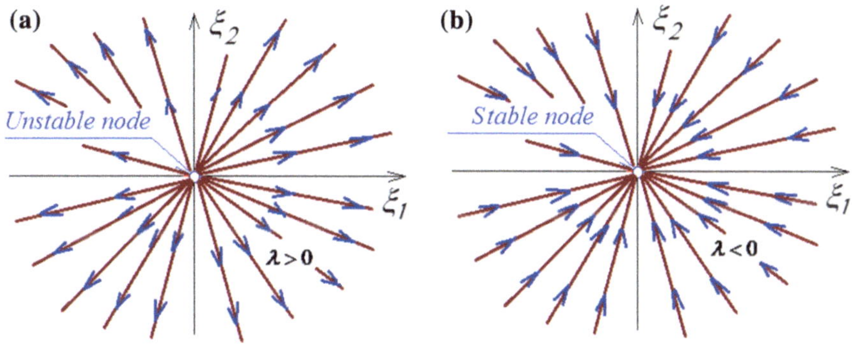

Fig. 7.9 Real, equal roots, the case II

It results after integration:

$$\xi_1 = \xi_{10}e^{\lambda t} \quad \text{and} \quad \xi_2 = \xi_{20}e^{\lambda t}. \tag{7.27}$$

By eliminating the time, trajectories as semi-straight lines beginning from origin are obtained. The critical (equilibrium) point is called the star and may be unstable (if $\lambda > 0$, Fig. 7.9a) or stable (if $\lambda < 0$, Fig. 7.9b).

(d) A root is real, different from zero and the second is zero.

The system of modal differential equations becomes:

$$\dot{\xi}_1 = \xi_1, \quad \dot{\xi}_2 = 0. \tag{7.28}$$

with the solutions:

$$\xi_1 = \xi_{10}e^{\lambda t}, \quad \xi_2 = \xi_{20}. \tag{7.29}$$

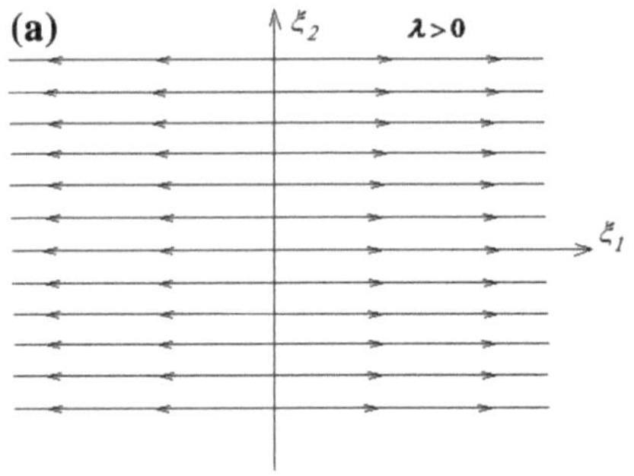

Fig. 7.10 One real root, the other zero

The trajectories will be semi-parallel to the axis ξ_1. There is a straight line of critical points in which the system is in neutral equilibria, namely the axis ξ_2. These points are stable if $\lambda < 0$ (Fig. 7.10a) and unstable if $\lambda > 0$ (Fig. 7.10b).

(e) The equation has a double root equal to zero. In this case there are two possible situations:

– The modal matrix is a second-order Jordan block:

$$\begin{bmatrix} 0 & 1 \\ 0 & 0 \end{bmatrix}. \tag{7.30}$$

The system of modal differential equations becomes:

$$\dot{\xi}_1 = \xi_2, \quad \dot{\xi}_2 = 0, \tag{7.31}$$

with the solutions:

$$\xi_1 = at + b, \quad \xi_2 = a. \tag{7.32}$$

The trajectories are parallel to the $O\xi_1$- axis, on each of them the speed is constant and proportional to ξ_2 (Fig. 7.11). The $O\xi_1$-axis is composed of critical points, which are unstable.

– The modal matrix is composed of two Jordan blocks of the first order:

$$\begin{bmatrix} 0 & 0 \\ 0 & 0 \end{bmatrix}. \tag{7.33}$$

The system of modal differential equations is, in this case:

$$\dot{\xi}_1 = 0, \quad \dot{\xi}_2 = 0, \tag{7.34}$$

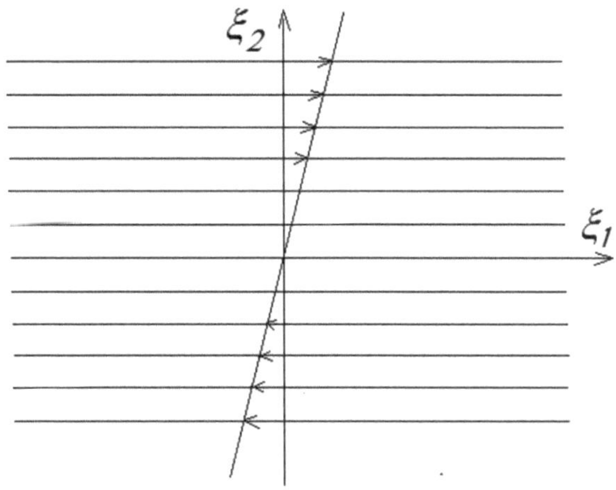

Fig. 7.11 Both roots are equal to zero

and has the solution:

$$\xi_1 = a, \quad \xi_2 = b. \tag{7.35}$$

In this case any point in the plane is a point of equilibrium. We say that we have a neutral equilibria.

7.3 Free Vibration of a Point

The previous analysis is made for a general system of two linear first-order differentiation equations. Let us now consider the free vibrations of a mechanical system with one degree of freedom. The equation of such a system is:

$$\ddot{x} + 2\beta\dot{x} + \omega_0^2 x = 0, \tag{7.36}$$

where $\beta > 0$ and ω_0^2 are the parameters that define the behavior of the material point. With the notation $\dot{x} = y$ the equation becomes:

$$\begin{aligned} \dot{x} &= y, \\ y &= -\omega_0^2 x - 2\beta y. \end{aligned} \tag{7.37}$$

The system matrix is:

$$\begin{bmatrix} 0 & 1 \\ -\omega_0^2 & -2\beta \end{bmatrix}, \tag{7.38}$$

and the characteristic equation:

$$\lambda^2 - 2\lambda\beta + \omega_0^2 = 0, \tag{7.39}$$

with the solutions:

$$\lambda_{1,2} = -\beta \pm \sqrt{\beta^2 - \omega_0^2}. \tag{7.40}$$

We identify the following cases:

(f) If $\beta^2 - \omega_0^2 > 0$, then the two roots λ_1 and λ_2 are real and negative and we are in case (b) previously studied in this section. The general solution is of the form:

$$x = C_1 e^{\lambda_1 t} + C_2 e^{\lambda_2 t}, \tag{7.41}$$

and, in a long enough time, becomes as small as possible.

(ii) If $\beta^2 - \omega_0^2 = 0$, we then have a real negative double root and the solution is (case f):

$$x = (C_1 + C_2 t)e^{-\beta t}, \tag{7.42}$$

and tends to zero if time tends to infinity.

(iii) If $\beta^2 - \omega_0^2 < 0$ (subcritical damping $\varsigma^2 = \left(\frac{\beta}{\omega_0}\right)^2 < 1$), then we have complex conjugated roots:

$$\lambda_{1,2} = -\beta \pm i\sqrt{\omega_0^2 - \beta^2}, \tag{7.43}$$

and the solution is:

$$x = Ae^{-\beta t}\cos(\omega t - \varphi), \tag{7.44}$$

where A and φ are integration constants (case d). By differentiation there is obtained:

$$\dot{x} = -\beta Ae^{-\beta t}\cos(\omega t - \varphi) - \omega Ae^{-\beta t}\sin(\omega t - \varphi) = -\beta x - \omega Ae^{-\beta t}\sin(\omega t - \varphi), \tag{7.45}$$

from here:

$$Ae^{-\beta t}\sin(\omega t - \varphi) = -\frac{\dot{x} + \beta x}{\omega}, \tag{7.46}$$

and:

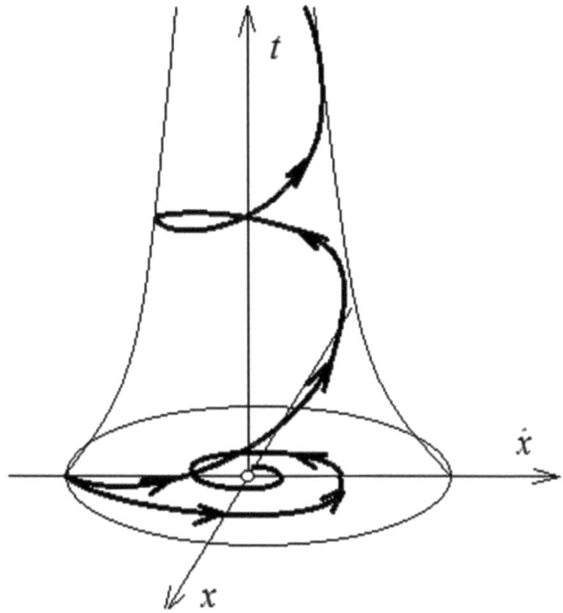

Fig. 7.12 Trajectories of a damped system

$$x^2 + \left(\frac{\dot{x} + \beta x}{\omega}\right)^2 = Ae^{-\beta t}. \tag{7.47}$$

The trajectory for this case is represented in Fig. 7.12.

If $\beta = 0$, the roots are purely imaginary, that is, of the form

$$\lambda_{1,2} = \pm i\omega_0, \tag{7.48}$$

and the solution is:

$$x = A\,\cos(\omega t - \varphi), \tag{7.49}$$

which means that we are in the case of (e). In this case, the system of first-order differential equations attached to the differential equation becomes, after some rearrangements, in homogeneous coordinates:

$$\begin{Bmatrix} \dot{x} \\ \frac{\dot{y}}{\omega_0} \end{Bmatrix} = \begin{bmatrix} 0 & \omega_0 \\ -\omega_0 & 0 \end{bmatrix} \begin{Bmatrix} x \\ \frac{y}{\omega_0} \end{Bmatrix}. \tag{7.50}$$

If we use the notation:

$$\{z\} = \begin{Bmatrix} x \\ \frac{y}{\omega_0} \end{Bmatrix}, \tag{7.51}$$

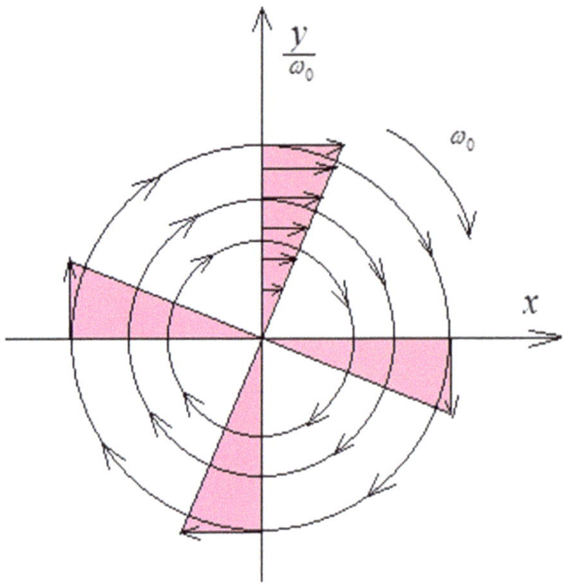

Fig. 7.13 Path in the homogeneous phase space

it results:

$$\{\dot{z}\} = \begin{bmatrix} 0 & \omega_0 \\ -\omega_0 & 0 \end{bmatrix} \{z\}. \tag{7.52}$$

This is formally equivalent to the relation defining the angular velocity of a rigid body with a rotation around a fixed axis:

$$\dot{\vec{r}} = \vec{\omega} x \vec{r}, \tag{7.53}$$

or, in matrix form:

$$\begin{Bmatrix} \dot{x} \\ \dot{y} \end{Bmatrix} = \begin{bmatrix} 0 & \omega_0 \\ -\omega_0 & 0 \end{bmatrix} \begin{Bmatrix} x \\ y \end{Bmatrix}. \tag{7.54}$$

In terms of homogeneous coordinates, the representation in the phase space is found in Fig. 7.13:

See Fig. 7.14.

Hence, by multiplying by $\{z\}^T$:

$$\{z\}^T \{\dot{z}\} = \{z\}^T \begin{bmatrix} 0 & \omega_0 \\ -\omega_0 & 0 \end{bmatrix} \{z\} = 0, \tag{7.55}$$

that is, $\{z\}$ and $\{\dot{z}\}$ are orthogonal. However:

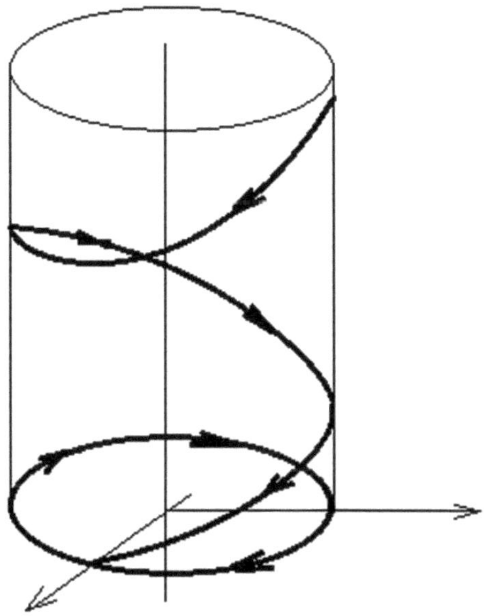

Fig. 7.14 Trajectory of the solutions without damping

$$\{z\}^T\{\dot{z}\} = \frac{d}{dt}\left(\frac{1}{2}\{z\}^T\{z\}\right), \tag{7.56}$$

thus

$$\frac{d}{dt}\left(\frac{1}{2}\{z\}^T\{z\}\right) = 0, \tag{7.57}$$

or

$$\frac{1}{2}\{z\}^T\{z\} = ct, \tag{7.58}$$

or

$$x^2 + \left(\frac{\dot{x}}{\omega_0}\right)^2 = \frac{kx^2 + m\dot{x}^2}{k} = \frac{2E}{m\omega_0^2}, \tag{7.59}$$

and represents the conservation of the total mechanical energy:

$$E = \frac{kx^2}{2} + \frac{m\dot{x}}{2}. \tag{7.60}$$

In the standard phase space the path given by Eq. (7.60) is an ellipse (see Fig. 7.15):

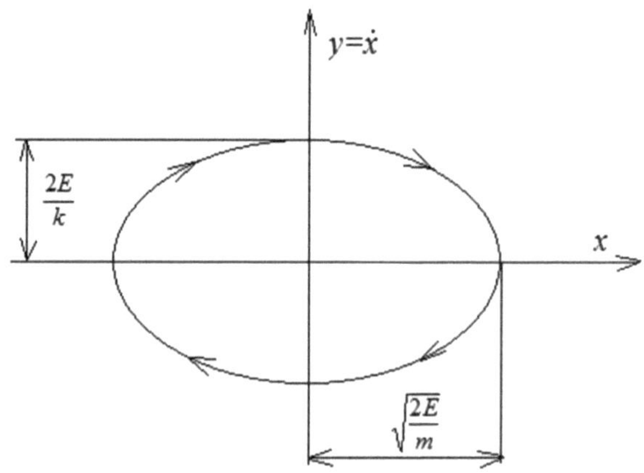

Fig. 7.15 Representation of the motion in the phase plane (conservation of mechanical energy)

$$\frac{x^2}{\left(\sqrt{\frac{2E}{m}}\right)^2} + \frac{\dot{x}^2}{\left(\sqrt{\frac{2E}{k}}\right)^2} = 1. \tag{7.61}$$

References

Guckenheimer, J., Holmes, P.H.: Nonlinear Oscillations Dynamical Systems and Bifurcations of Vector Fields. Springer, New-York, Berlin, Heilderberg, Tokyo (1983)

Alligood, K.T., Sauer, T.D., Yorke, J.A., Chaos: An Introduction to Dynamical Systems. Springer, New York (2000)

Anosov, D.V., Arnold, V.I.: Dynamical Systems. Springer, Berlin, Heildelberg, New York, London, Paris, Tokyo (1988)

Arnold, V.I.: Ordinary differential equations. Ed. In: Stiintifica si Enciclopedica. Bucharest (1973)

Arnold, V.I.: Mathematical Methods of Classical Mechanics. Springer, New York (2013)

MIX
Papier aus verantwortungsvollen Quellen
Paper from responsible sources
FSC® C105338

If you have any concerns about our products,
you can contact us on
ProductSafety@springernature.com

In case Publisher is established outside the EU,
the EU authorized representative is:
Springer Nature Customer Service Center GmbH
Europaplatz 3, 69115 Heidelberg, Germany

Printed by Libri Plureos GmbH
in Hamburg, Germany